Adapting to the Impacts of Climate Change

America's Climate Choices:
Panel on Adapting to the Impacts of Climate Change

Board on Atmospheric Sciences and Climate

Division on Earth and Life Studies

NATIONAL RESEARCH COUNCIL
OF THE NATIONAL ACADEMIES

THE NATIONAL ACADEMIES PRESS
Washington, D.C.
www.nap.edu

THE NATIONAL ACADEMIES PRESS · 500 Fifth Street, N.W. · Washington, DC 20001

NOTICE: The project that is the subject of this report was approved by the Governing Board of the National Research Council, whose members are drawn from the councils of the National Academy of Sciences, the National Academy of Engineering, and the Institute of Medicine. The members of the committee responsible for the report were chosen for their special competences and with regard for appropriate balance.

This study was supported by the National Oceanic and Atmospheric Administration under contract number DG133R08CQ0062. Any opinions, findings, conclusions, or recommendations expressed in this material are those of the author(s) and do not necessarily reflect the views of the sponsoring agency or any of its sub agencies.

International Standard Book Number-13:   978-0-309-14591-6 (Book)
International Standard Book Number-10:   0-309-14591-0 (Book)
International Standard Book Number-13:   978-0-309-14592-3 (PDF)
International Standard Book Number-10:   0-309-14592-9 (PDF)
Library of Congress Control Number: 2010940139

Additional copies of this report are available from the National Academies Press, 500 Fifth Street, N.W., Lockbox 285, Washington, DC 20055; (800) 624-6242 or (202) 334-3313 (in the Washington metropolitan area); Internet, http://www.nap.edu

Cover images:

Far left: courtesy of NASA/GSFC/METI/ERSDAC/JAROS, and U.S./Japan ASTER Science Team

Middle left: courtesy of U.S. Department of Agriculture Natural Resources Conservation Service

Middle right: courtesy of Fotosearch Stock Photography

Far right: courtesy of University Corporation for Atmospheric Research, Photo by Carlye Calvin

Printed in the United States of America

THE NATIONAL ACADEMIES
Advisers to the Nation on Science, Engineering, and Medicine

The **National Academy of Sciences** is a private, nonprofit, self-perpetuating society of distinguished scholars engaged in scientific and engineering research, dedicated to the furtherance of science and technology and to their use for the general welfare. Upon the authority of the charter granted to it by the Congress in 1863, the Academy has a mandate that requires it to advise the federal government on scientific and technical matters. Dr. Ralph J. Cicerone is president of the National Academy of Sciences.

The **National Academy of Engineering** was established in 1964, under the charter of the National Academy of Sciences, as a parallel organization of outstanding engineers. It is autonomous in its administration and in the selection of its members, sharing with the National Academy of Sciences the responsibility for advising the federal government. The National Academy of Engineering also sponsors engineering programs aimed at meeting national needs, encourages education and research, and recognizes the superior achievements of engineers. Dr. Charles M. Vest is president of the National Academy of Engineering.

The **Institute of Medicine** was established in 1970 by the National Academy of Sciences to secure the services of eminent members of appropriate professions in the examination of policy matters pertaining to the health of the public. The Institute acts under the responsibility given to the National Academy of Sciences by its congressional charter to be an adviser to the federal government and, upon its own initiative, to identify issues of medical care, research, and education. Dr. Harvey V. Fineberg is president of the Institute of Medicine.

The **National Research Council** was organized by the National Academy of Sciences in 1916 to associate the broad community of science and technology with the Academy's purposes of furthering knowledge and advising the federal government. Functioning in accordance with general policies determined by the Academy, the Council has become the principal operating agency of both the National Academy of Sciences and the National Academy of Engineering in providing services to the government, the public, and the scientific and engineering communities. The Council is administered jointly by both Academies and the Institute of Medicine. Dr. Ralph J. Cicerone and Dr. Charles M. Vest are chair and vice chair, respectively, of the National Research Council.

www.national-academies.org

PANEL ON ADAPTING TO THE IMPACTS OF CLIMATE CHANGE

KATHARINE L. JACOBS* (*Chair—through January 3, 2010*), University of Arizona, Tucson

THOMAS J. WILBANKS (*Chair*), Oak Ridge National Laboratory, Tennessee

BRUCE P. BAUGHMAN, IEM, Inc., Alabaster, Alabama

ROGER N. BEACHY,* Donald Danforth Plant Science Center, St. Louis, Missouri

GEORGES C. BENJAMIN, American Public Health Association, Washington, D.C.

JAMES L. BUIZER, Arizona State University, Tempe

F. STUART CHAPIN III, University of Alaska, Fairbanks

W. PETER CHERRY, Science Applications International Corporation, Ann Arbor, Michigan

BRAXTON DAVIS, South Carolina Department of Health and Environmental Control, Charleston

KRISTIE L. EBI, IPCC Technical Support Unit WGII, Stanford, California

JEREMY HARRIS, Sustainable Cities Institute, Honolulu, Hawaii

ROBERT W. KATES, Independent Scholar, Bangor, Maine

HOWARD C. KUNREUTHER, University of Pennsylvania Wharton School of Business, Philadelphia

LINDA O. MEARNS, National Center for Atmospheric Research, Boulder, Colorado

PHILIP MOTE, Oregon State University, Corvallis

ANDREW A. ROSENBERG, Conservation International, Arlington, Virginia

HENRY G. SCHWARTZ, JR., Jacobs Civil (retired), St. Louis, Missouri

JOEL B. SMITH, Stratus Consulting, Inc., Boulder, Colorado

GARY W. YOHE, Wesleyan University, Middletown, Connecticut

NRC Staff

CLAUDIA MENGELT, Study Director
MICHAEL CRAGHAN, Program Officer
KARA LANEY, Associate Program Officer
JOSEPH CASOLA, Postdoctoral Fellow
LAUREN M. BROWN, Research Associate
AMANDA PURCELL, Senior Program Assistant

---

* Asterisks denote members who resigned during the study process.

Foreword: About America's Climate Choices

Convened by the National Research Council in response to a request from Congress (P.L. 110-161), *America's Climate Choices* is a suite of five coordinated activities designed to study the serious and sweeping issues associated with global climate change, including the science and technology challenges involved, and to provide advice on the most effective steps and most promising strategies that can be taken to respond.

The Committee on America's Climate Choices is responsible for providing overall direction, coordination, and Integration of the *America's Climate Choices* suite of activities and ensuring that these activities provide well-supported, action-oriented, and useful advice to the nation. The committee convened a Summit on America's Climate Choices on March 30–31, 2009, to help frame the study, and provide an opportunity for high-level participation and input on key issues. The committee is also charged with writing a final report that builds on four panel reports and other sources to answer the following four overarching questions:

- What short-term actions can be taken to respond effectively to climate change?
- What promising long-term strategies, investments, and opportunities could be pursued to respond to climate change?
- What are the major scientific and technological advances needed to better understand and respond to climate change?
- What are the major impediments (e.g., practical, institutional, economic, ethical, intergenerational, etc.) to responding effectively to climate change, and what can be done to overcome these impediments?

The Panel on Limiting the Magnitude of Future Climate Change was charged to describe, analyze, and assess strategies for reducing the net future human influence on climate. The panel's report focuses on actions to reduce domestic greenhouse gas emissions and other human drivers of climate change, such as changes in land use, but also considers the international dimensions of climate stabilization.

The Panel on Adapting to the Impacts of Climate Change was charged to describe, analyze, and assess actions and strategies to reduce vulnerability, increase adaptive capacity, improve resiliency, and promote successful adaptation to climate change in different regions, sectors, systems, and populations. This report draws on a wide range of sources and case studies to identify lessons learned from past experiences, promising current approaches, and potential new directions.

The Panel on Advancing the Science of Climate Change was charged to provide a concise overview of past, present, and future climate change, including its causes and its impacts, and to recommend steps to advance our current understanding, including new observations, research programs, next-generation models, and the physical and human assets needed to support these and other activities. The panel's report focuses on the scientific advances needed both to improve our understanding of the intergrated human-climate system and to devise more effective responses to climate change.

The Panel on Informing Effective Decisions and Actions Related to Climate Change was charged to describe and assess different activities, products, strategies, and tools for informing decision makers about climate change and helping them plan and execute effective, integrated responses. The panel's report describes the different types of climate change-related decisions and actions being taken at various levels and in different sectors and regions; it develops a framework, tools, and practical advice for ensuring that the best available technical knowledge about climate change is used to inform these decisions and actions.

America's Climate Choices builds on an extensive foundation of previous and ongoing work, including National Research Council reports, assessments from other national and international organizations, the current scientific literature, climate action plans by various entities, and other sources. More than a dozen boards and standing committees of the National Research Council were involved in developing the study, and many additional groups and individuals provided additional input during the study process. Outside viewpoints were also obtained via public events and workshops (including the Summit), invited presentations at committee and panel meetings, and comments and questions received through the study website, *http://americasclimate-choices.org*.

Collectively, the *America's Climate Choices* suite of activities involves more than 90 volunteers from a range of communities including academia, various levels of government, business and industry, other nongovernmental organizations, and the international community. Responsibility for the final content of each report rests solely with the authoring panel and the National Research Council. However, the development of each report included input from and interactions with members of all five study groups; the membership of each group is listed in Appendix A.

Preface

This report presents the findings of the Committee on Adapting to the Impacts of Climate Change, one of four concurrent panel efforts within the *America's Climate Choices* committee study. It was our assignment to identify the opportunities and challenges associated with adaptation, to identify and evaluate the available options and lessons learned within the United States and elsewhere, and to make recommendations regarding U.S. adaptation efforts.

Adapting to climate change is a relatively new topic for U.S. citizens, who have only recently become fully aware of the implications of changes in the Earth system that will result from having more heat trapped in the oceans and the atmosphere. In recent years, some states, cities, and sectors have begun to make plans to adapt to current and anticipated changes in the climate system. Some "early adopters" have focused primarily on limiting greenhouse gases (GHGs). Others, however, are also addressing ways to limit impacts of the anticipated changes, recognizing that regardless of efforts to limit emissions, adaptation is required now and will become even more important in the coming decades. Although planning for adaptation is still in its infancy, there is a groundswell of interest in moving forward quickly to avoid future impacts of climate change.

Advising the nation on how to prepare for the impacts of climate change is especially daunting in a country with so much geographic and economic diversity and so many private- and public-sector decision makers. The challenges associated with multiple regions, sectors, scales, and time frames have made this a difficult assignment, and in the end, our panel has concluded that is not possible to provide a list of actions to be taken now to adapt in each region and sector. As has been noted by many researchers and practitioners, adaptation is fundamentally implemented at local and regional levels and needs to consider the socioeconomic and political factors. Priorities regarding "what to do" need to be set in decision contexts relative to other important priorities faced by society and resource managers. Vulnerability associated with climate change is based on underlying social and ecological stresses, and these stresses tend to vary dramatically from place to place. Degrees of vulnerability are not directly connected to wealth, but certainly a lack of financial capacity is highly correlated with a reduced number of options for adaptation. In this report, our panel emphasizes that adaptation decisions need to be made in the context of promoting long-term sustainability objectives, including social, economic, and ecological welfare rather than focusing only on the short-term outcomes that may be more politically and economically expedient.

Despite this place-based framework, our panel shares the perspective that adaptation needs to be addressed in a coordinated way and that there is a need to involve the federal government in this coordination. Furthermore, there is a need to acknowledge the implications of our adaptation and GHG-reduction decisions on national security and to be prepared for the potential impacts of decisions taken by other countries.

This assignment has been both challenging and exhilarating for other reasons as well. Although dozens of new publications on adaptation have emerged during the year that we have worked on this effort, on balance there is very little published literature about the effectiveness of alternative approaches to adaptation to impacts of climate change, and in particular very few estimates of cost that are useful in the context of the wide variety of U.S. decision processes. The exhilarating part of this effort has been the opportunity to meld a variety of kinds of knowledge into a truly integrated document that benefits from a balance between social and physical science and practical experience.

We were aided in our efforts by the support of truly exceptional National Research Council staff. Our project director, Claudia Mengelt, did a heroic job at maintaining forward momentum and managing this intensive effort. She was unceasingly energetic, professional, and optimistic, in spite of relatively severe time limitations and a large committee of talented but very independent-minded members. Claudia was ably assisted in her work by Amanda Purcell, who impeccably handled the logistics; Michael Craghan, who developed the matrix format and did much of the citation development; and Kara Laney, who assisted with our research in multiple ways. We also want to acknowledge the highly professional stewardship provided by the study director, Ian Kraucunas, and the Board on Atmospheric Sciences and Climate (BASC) board director, Chris Elfring, who engaged often with our committee to provide sound advice, particularly about coordination with the other committee and panel findings. We depended heavily on the U.S. Global Change Research Program report *The Impacts of Climate Change on the United States,* and are grateful to those who helped to produce it and shared their findings with us firsthand. The *Impacts* report informed our conclusions about what climate impacts we need to be prepared for. Many international and national climate and adaptation experts shared their expertise with us in person, by phone, and through documents they provided. Their input was invaluable and used liberally in the case studies and findings of this report.

Our committee is grateful for the opportunity to work together at this important moment in history, when climate change science and policy are intersecting for the first time as part of a major national agenda. We are humbled by the size of this task and the magnitude of the known and unknown challenges that lie ahead, especially on the ambitious time schedule for the *America's Climate Choices* reports.

Katharine Jacobs, Chair through January 3, 2010, and *Tom Wilbanks*, Chair
Panel on Adapting to the Impacts of Climate Change

Acknowledgments

This report has been reviewed in draft form by individuals chosen for their diverse perspectives and technical expertise, in accordance with procedures approved by the NRC's Report Review Committee. The purpose of this independent review is to provide candid and critical comments that will assist the institution in making its published report as sound as possible and to ensure that the report meets institutional standards for objectivity, evidence, and responsiveness to the study charge. The review comments and draft manuscript remain confidential to protect the integrity of the deliberative process. We wish to thank the following individuals for their participation in the review of this report:

NEIL ADGER, Tyndall Centre for Climate Change Research, Norwich, U.K.

DONALD F. BOESCH, University of Maryland Center for Environmental Science, Cambridge

IAN BURTON, Meteorological Service of Canada, Ontario

JONATHAN CANNON, University of Virginia, Charlottesville

MARGARET DAVIDSON, National Oceanic and Atmospheric Administration, Charleston, South Carolina

ALEXANDER H. FLAX, Consultant, Potomac, Maryland

AMY FRAENKEL, UNEP Regional Office for North America, Washington, D.C.

GERALD E. GALLOWAY, University of Maryland, College Park

JAMES E. GERINGER, Environmental Systems Research Institute, Cheyenne, Wyoming

GEORGE M. HORNBERGER, Vanderbilt University, Nashville, Tennessee

PETER KAREIVA, The Nature Conservancy, Seattle, Washington

JIM LOPEZ, U.S. Department of Housing and Urban Development, Washington, D.C.

RICHARD H. MOSS, Pacific Northwest National Laboratory, Washington, D.C.

DAVID J. NASH, Dave Nash & Associates, LLC, Birmingham, Alabama

CHARLES PHELPS, University of Rochester, Gualala, California

KEITH PITTS, Marrone Organic Innovations, Davis, California

PEGGY M. SHEPARD, WE ACT for Environmental Justice, New York, New York

B. L. TURNER II, Arizona State University, Tempe

Although the reviewers listed above have provided many constructive comments and suggestions, they were not asked to endorse the conclusions or recommendations nor did they see the final draft of the report before its release. The review of this report was overseen by Robert A. Frosch (Harvard University) and Susan Hanson (Clark University) appointed by the Report Review Committee and the Division on Earth and Life Studies, who were responsible for making certain that an independent examination of this report was carried out in accordance with institutional procedures and that all review comments were carefully considered. Responsibility for the final content of this report rests entirely with the authoring committee and the institution.

The panel would like to thank in particular the following for sharing their insights on this topic as presenters and informal reviewers and for writing contributions: Thomas Armstrong, Peter Schultz, Joel Scheraga, Susan Solomon, Jerry Melillo, Brad Udall, Dixon Butler, Jean Fruci, Kris Sarri, Susanne Moser, Amanda Staudt, Saleemul Huq, Mark Howden, Chris West, Virginia Burkett, Michael Savonis, Matthias Ruth, Adam Freed, Tony Brunello, Mark Way, Andrew Castaldi, Hal Mooney, Lisa Graumlich, Peter Culp, Jennifer Pitt, Nancy Grimm, Mikaela Engert, Mitzi Stults, Jim Jones, and John Reilly.

Institutional oversight for this project was provided by:

ACKNOWLEDGMENTS

NRC Staff

CHRIS ELFRING, Director
LAURIE GELLER, Senior Program Officer
IAN KRAUCUNAS, Senior Program Officer
EDWARD DUNLEA, Senior Program Officer
MARTHA MCCONNELL, Program Officer
TOBY WARDEN, Program Officer
MAGGIE WALSER, Associate Program Officer
KATIE WELLER, Associate Program Officer
JOSEPH CASOLA, Postdoctoral Fellow
RITA GASKINS, Administrative Coordinator
LAUREN M. BROWN, Research Associate
ROB GREENWAY, Program Associate
SHELLY FREELAND, Senior Program Assistant
AMANDA PURCELL, Senior Program Assistant
RICARDO PAYNE, Senior Program Assistant
JANEISE STURDIVANT, Program Assistant
SHUBHA BANSKOTA, Financial Associate

Contents

Summary

The global climate is changing, and impacts of climate change are being observed across the United States. Over the past 50 years, temperatures have risen nearly 2°F (1°C), some extreme weather events such as heavy precipitation and heat waves have increased in frequency and intensity, sea level has risen along most of the coast, and sea ice has been disappearing rapidly. These changes are all expected to continue, which means that in many respects the climate of the future will be different from the climate of the past.

In order to address the challenges associated with climate change, Congress directed the National Research Council to "investigate and study the serious and sweeping issues relating to global climate change and make recommendations regarding the steps that must be taken and what strategies must be adopted in response to global climate change." As part of the response to this request, the America's Climate Choices (ACC) Panel on Adapting to the Impacts of Climate Change was charged to "describe, analyze, and assess actions and strategies to reduce vulnerabilities, increase adaptive capacity, improve resilience, and promote successful adaptation to climate change in different regions, sectors, systems, and populations" (see Appendix B for the full statement of task).

America's climate change *adaptation* choices involve deciding how to cope with climate changes that we cannot, or do not, avoid so that possible disruptions and damages to society, economies, and the environment are minimized and—where possible—so that impacts are converted into opportunities for the country and its citizens. In some cases, such as in Alaska, the need to adapt has already become a reality. In most cases, however, adapting today is about reducing vulnerabilities to emerging or future impacts that could become seriously disruptive if we do not begin to identify response options now; in other words, adaptation today is essentially a risk management strategy.

Vulnerabilities to climate change impacts exist all across America and differ by region, sector, scale, and segment of our society. Consider, for example, the likelihood of reduced surface water supply in America's West because of reduced snowfall and snowpack in the western mountains and, at least in the Southwest, prospects for reduced total rainfall. These changes interact with the region's current vulnerabilities to drought conditions and the many competing demands for limited water resources.

Options for adapting to the prospect of more severe water shortage in the West and Southwest include improving efficiencies in water use, reducing the need for water for competing purposes (e.g., power plant cooling), finding ways to reduce evaporation from reservoirs, learning more about potentials and limits of groundwater withdrawal, increasing mechanisms for interbasin water transfers, revisiting approaches to water rights, and developing technology for affordable desalination of sea water. These are examples of options that can be considered by decision makers responsible for water resources in the context of the local or regional socioeconomics, combining relatively low-cost near-term actions with preparations to evaluate more substantial actions in the longer term. While it is difficult to know precisely the impacts that will occur in the future, adaptation offers a way to prepare and minimize the risks to social, economic, and natural systems associated with these impacts.

Adaptation to reduce vulnerabilities associated with likely impacts of climate change cannot be accomplished by the federal government or any other single decision maker alone. The challenges are too diverse, the contexts are too different, and too many parties have knowledge and capacities to contribute. Given the diversity of climate impacts, vulnerabilities, and available adaptation options across the United States, the report concludes that adaptation planning and action will be required across all levels of government as well as within the private sector, nongovernmental organizations (NGOs), and community organizations. Accordingly, this report outlines a framework that engages decision makers across all levels of governance and across public and private entities through the development of a national adaptation strategy. Within this national strategy, the federal government plays a unique and critical role in providing technical and scientific resources that are lacking at the local or regional scale, reexamining policies that may inhibit adaptation, and supporting scientific research to expand our knowledge of impacts and adaptation.

FUTURE IMPACTS OF CLIMATE CHANGE THAT CALL FOR ADAPTATION

Effective adaptation depends on an understanding of projected climatic changes at geographic and temporal scales appropriate for the needed response. Projected changes include average and extreme temperature; average and extreme precipitation; the intensity, frequency, duration, and/or location of extreme weather events; sea level; and atmospheric carbon dioxide (CO_2) concentrations. Because of the complex interactions between these climate changes and nonclimate factors, such as demographics, economics, land use, and technology, the impacts of climate change will be highly diverse. For example, future climate changes will interact with underlying vulnerabilities in many coastal communities. In areas that have been highly developed, the ability to cope with flooding has been reduced as wetlands have been drained. With projected sea level rise and increases in storm surge, the impacts of flood damage and coastal erosion could be exacerbated. Thus, effective approaches to adaptation will be case- and place-specific.

Society's ability to cope with the impacts of climate change and avoid unacceptable levels of social and environmental costs decreases as the severity of climate change increases. At moderate rates and levels of climate change, adaptation can do a great deal. At severe rates and levels of climate change, however, limits of many adaptation options might be reached; resulting adaptations are likely to be much more disruptive and costly. In this very direct and profound sense, adaptation to the impacts of climate change and actions to reduce greenhouse emissions into the atmosphere are partners in America's response to concerns about climate change, not alternatives.

Many scientific challenges remain in assessing vulnerabilities and impacts associated with climate change. The level of scientific confidence in understanding and projecting climate change increases with increasing spatial scale while the relevance and value of the information to decision makers declines. Therefore, a finer-scale understanding of climate change risks and vulnerabilities is needed. In addition, multiple stresses will interact with climate change in determining its impacts and, because vulnerability varies greatly from place to place, the same climate condition in different locations may call for different adaptive responses.

OPTIONS FOR ADAPTING TO IMPACTS OF CLIMATE CHANGE

If the United States is to cope effectively with the impacts of climate change, it will need an array of adaptation options from which to choose. Until very recently, adapting to climate change has been a low national priority, and limited research has been completed to identify options for adaptation and evaluate their benefits, costs, potential, and limits. In the short term, the nation can draw lessons from past experience with adaptations to climate *variability*, experience (albeit limited) with climate change adaptation that has been undertaken in some regions of the world, a limited number of careful analyses of adaptation possibilities, and an onrush of creative thinking in connection with emerging efforts to do adaptation planning. But, in many cases, the options that we can identify for *adaptation to impacts of climate change* lack solid information about benefits, costs, potentials, and limits for three reasons: an inability to attribute explicitly many observed changes at local and regional scales to climate change (and therefore to document effects of adaptation in reducing those impacts), the diversity of impacts and vulnerabilities across the United States, and the relatively small body of research that focuses on climate change adaptation actions.

This report provides examples of the range of options currently available for adapting to climate variability and extremes in key climate-sensitive sectors, such as agriculture, energy, and transportation. Although these examples alone may not be sufficient for coping with future climate change, they offer a starting point for devising adaptation strategies. While the report provides a long list of options to be considered for various sectors, Table S.1 illustrates the range and diversity of options for coastal regions. For example, options to cope with sea level rise near coastal areas include hardening of coastal infrastructure so that it can handle higher water levels and storm impacts, sharing risks among vulnerable locations through insurance, and altering development and land-use practices to relocate vulnerable infrastructure or activities away from the coasts. Some of the adaptation options can be implemented in the near term at relatively low cost or provide additional benefits. Early actions that can be deployed most easily in such an environment are likely to be low-cost strategies with win-win outcomes, actions that end or reverse maladapted policies and practices, and measures that avoid prematurely narrowing future adaptation options. In addition, the integration of efforts to reduce greenhouse gas (GHG) emissions and adapt to climate change impacts in a common sustainability agenda reduces risks of maladaptation.

In the long term, adaptation to climate change calls for a new paradigm that takes into account a range of possible future climate conditions and associated changes in human and natural systems instead of managing our resources based on previous experience and the historical range and variability of climate. This does not mean waiting until uncertainties have been reduced to consider adaptation actions. Actions taken now can reduce the risk of major disruptions to human and natural systems; inaction could serve to increase these risks, especially if the rate or magnitude of climate change is particularly large. Mobilizing now to increase the nation's adaptive capacity can be viewed as an insurance policy against an uncertain future. Because adaptation options are much more limited to cope with impacts of relatively severe climate change in the longer run, an important part of a national approach to adaptation is examining the prospects for these more severe impacts and considering possible limits to adaptation. Some projected impacts are likely to be beyond the scope of adaptation, unless adaptation involves major structural change to government and society.

DEVELOPING ADAPTATION STRATEGIES

Although many ideas are available about ways to adapt to climate variability and change, few of these options have been assessed for their effectiveness under projected future climate conditions and for their potential interactions across sectors and with other stressors. Little attention has been given to the processes that decision makers might use to make appropriate adaptation decisions. This report suggests some approaches to choosing among the many options to manage the risks associated with climate change, using examples from recent adaptation activities initiated primarily at the state and local levels.

In brief, the report suggests that the adaptation process is fundamentally a risk-management strategy. Managing risk in the context of adapting to climate change involves using the best available social and physical science to understand the likelihood of climate impacts and their associated consequences, then selecting and implementing the response options that seem most effective. Because knowledge about future impacts and the effectiveness of response options will evolve, policy decisions to manage the risk of climate change impacts can be improved if they are done in an iterative fashion by continually monitoring the progress and consequences of actions and modifying management practices based on learning and recognition of changing conditions.

The report proposes a sequence of steps for pursuing adaptation. To begin, decision makers across a variety of agencies and institutions (e.g., federal, tribal, state, and local governments; private-sector firms; and community organizations and NGOs) would identify their vulnerabilities and assess risks associated with the impacts of climate change. This information would need to be communicated among stakeholders and relevant decision makers to raise their awareness of current and potential problems. Using a risk management approach, adaptation options for managing the risks associated with climate impacts can then be identified, evaluated, and implemented (Figure S.1).

The report also identifies some "lessons learned" about important elements to developing a strategy, including establishment of clear objectives, opportunities to incorporate adaptation plans into existing management goals and procedures, the ability to identify co-benefits associated with adaptation measures, and the presence of strong leadership.

TABLE S.1 Possible options for adapting to climate change that have been identified in the ocean and coastal sector.

Climate Change	Impact	Possible Adaptation Action	Federal	State	Local Government	Private Sector	NGO/Individuals
Accelerated sea level rise and lake level changes	Gradual inundation of low-lying land; loss of coastal habitats, especially coastal wetlands; saltwater intrusion into coastal aquifers and rivers; increased shoreline erosion and loss of barrier islands; changes in navigational conditions	Site and design all future public works projects to take into account projections for sea level rise.	■	■	■		
		Eliminate public subsidies for future development in high hazard areas along the coast.	■	■			
		Develop strong, well-planned, shoreline retreat or relocation plans and programs (public infrastructure and private properties), and poststorm redevelopment plans.		■	■		
		Retrofit and protect public infrastructure (stormwater and wastewater systems, energy facilities, roads, causeways, ports, bridges, etc.).	■	■	■		
		Adapt infrastructure and dredging to cope with altered water levels.	■	■			
		Use natural shorelines, setbacks, and buffer zones to allow inland migration of shore habitats and barrier islands over time (e.g., dunes and forested buffers mitigate storm damage and erosion).	■	■	■		■
		Encourage alternatives to shoreline "armoring" through "living shorelines" (NRC).	■	■	■		
		Develop strategic property acquisition programs to discourage development in hazardous areas, encourage relocation, and/or allow for inland migration of intertidal habitats.	■	■			

Impact	Changes	Response strategy					
Changes in sea ice	Changes in ecosystem structures	Plan and manage ecosystems to encourage adaptation (see ecosystem options).		■	■		
	Exacerbate coastal erosion; severe storms reach coast	Facilitate inland migration and relocation of coastal communities.	■	■	■	■	
Increased intensity/frequency coastal storms	Increased storm surge and flooding; increased wind damage; sudden coastal/shoreline alterations	Strengthen and implement building codes that make existing buildings more resilient to storm damage along the coast.	■	■			
		Increase building "free board" above base flood elevation	■	■	■		
		Identify and improve evacuation routes in low-lying areas (e.g., causeways to coastal islands).	■	■	■		
		Improve storm readiness for harbors and marinas.	■	■	■	■	
		Establish marine debris reduction strategy.	■	■	■	■	
		Establish and enforce shoreline setback requirements.	■	■			
Ocean acidification	Potential changes in ocean productivity and food web linkages; degradation of corals, shellfish, and other shelled organisms; potential impacts on coastal infrastructure (i.e., construction materials)	Reduce CO_2 emissions (Limiting).	■	■	■	■	
		Support ocean observation and long-term monitoring programs.	■	■	■		
		Evaluate and manage for ecosystem and infrastructure impacts.					■

7

Category	Action											
Changes in physical and chemical characteristics of marine systems Changes in salinity; changes in circulation; changes in seawater temperature; changes in salinity and temperature stratification; changes in estuarine structure and processes (e.g., salt wedge migration); changes in ecosystem structure ("invasive," nonnative species), species distributions, population genetics, and life history strategies (including migratory routes for protected and commercially important species); increased frequency and extent of harmful algal blooms and coastal hypoxia events	Establish monitoring and mapping efforts to measure changes in physical, biological, and chemical conditions along the coast.	■	■				■					■
	Utilize approaches that do not endanger species that are harvested or endangered.	■	■				■					■
	Ensure flexibility in management plans to account for changes in species distributions and abundance.						■			■		■
	Implement early warning and notification systems for shellfish and beach closures, salinity intrusion in coastal rivers (for industry impacts and water resource management, i.e. freshwater intakes), and for unusual events such as hypoxia.	■	■	■			■		■	■		■
Changes in precipitation Increased runoff and non-point source pollution or eutrophication; changes in coastal hydrology and related ecosystem impacts; increased coastal flooding	Improve non-point source pollution prevention programs.	■	■				■		■	■		■
	Improve stormwater management systems and infrastructure.								■	■		
	Improve early warning systems for beach and shellfish closures.	■	■				■			■		■

NOTE: Most adaptations are local and need to be tailored to local conditions. The suitability of each adaptation option must therefore be evaluated in the context of local conditions. Where possible, the table refers to assessments and syntheses that consider multiple adaptation options and provide references to specific studies.

SOURCE: Reference citations are abbreviated as follows to conserve space: NRC (NRC, 2007c), Limiting (ACC: *Limiting the Magnitude of Future Climate Change* [NRC, 2010c]).

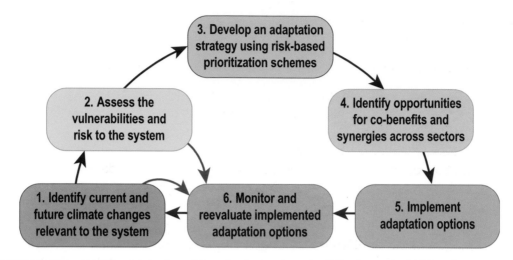

FIGURE S.1 The planning process is envisioned to incorporate the following steps: (1) identify current and future climate changes relevant to the system, (2) assess the vulnerabilities and risk to the system, (3) develop an adaptation strategy using risk-based prioritization schemes, (4) identify opportunities for co-benefits and synergies across sectors, (5) implement adaptation options, and (6) monitor and reevaluate implemented adaptation options.

LINKING ADAPTATION EFFORTS ACROSS THE NATION

Adapting to climate change impacts is and will be an ongoing process. It cannot be thought of simply as a set of actions to be taken right now, although this report does identify some effective short-term actions. Adapting calls for the development of a multiparty, public-private national framework for becoming more adaptable over time, including improving information systems for telling us what is happening, both with climate change impacts and with adaptation experiences; working together across institutional and social boundaries to combine what each party does best; and making it a part of our national culture to continually review the effectiveness of current risk-management strategies as we learn more about projected climate changes and impact vulnerabilities.

In this sense, adaptation poses enormous challenges across sectors, jurisdictions, and levels of governance. Successful adaptation to climate change involves a multitude of interested partners and decision makers: federal, state, and local governments; the private sector, large and small; NGOs and community groups; and others. The issue is how to create a framework in which all of the parties work together effectively, taking advantage of the strengths of each and assuring that the activities reinforce each other rather than getting in each other's way.

There are three general kinds of alternative approaches for meeting this need:

1. A strong federal government adaptation program, nested in a body of federal government laws, regulations, and institutions. With this approach, the federal government would take the lead in identifying adaptation actions in the national interest, mandate appropriate responses while providing resources to support them, set goals for improvements in the nation's adaptive capacities, and ensure coordination with other national programs and parties nationwide.

2. A grassroots-based, bottom-up approach that is very largely self-driven. Adaptation planning and actions would be decentralized. Decisions would be made without significant federal encouragement or coordination, except for programs of the federal agencies themselves. Current adaptation efforts are largely occurring in this manner.

3. An intermediate approach, where planning and actions are decentralized but the federal government plays a significant role as a catalyst and coordinator at the outset, providing information and technical resources and continually evaluating needs for additional risk management at a national level.

The panel considered all three approaches, in consultation with social scientists, practitioners, and stakeholders, and found that the intermediate approach was the alternative with the strongest scientific support, because adaptation requires place-based approaches in combination with technical and scientific capacity typically developed at the federal level. Based on its review of recent reports and in consultation with stakeholders, the panel also concludes that practitioners and stakeholders favor the intermediate approach. Elaborating on this approach, the panel found that emerging adaptation efforts in the United States are not well coordinated, and as a result adaptation choices could result in unintended consequences and inconsistent, inefficient investments and outcomes. A national adaptation program is needed, guided by a strategy that focuses on cooperation and collaboration among different levels of government and between government and other key parties.

A national adaptation program itself will need to be adaptive, continually working to increase its own effectiveness. Solutions need to be developed that promote response to changing conditions, informed by ongoing information collection and dissemination, as opposed to a rigid response intended to be permanent. An ongoing assessment of progress (in terms of both outcomes and process) is an integral part to the success of this program. Other critical features of adaptive management involve learning from past and emerging experiences, recognizing the complexity and the interrelated nature of sectoral interests such as water, agriculture, and energy, and understanding the relationships between adaptation activities and the need to limit GHG emissions. Over time, there will be a need to adapt to our own adaptations (and maladaptations) as well as to our efforts to limit the magnitude of climate change.

THE INTERNATIONAL CONTEXT FOR AMERICA'S ADAPTATION EFFORTS

Engaging in international dialogues and actions about climate change adaptation could have several benefits for the United States. First, it would help address questions of global equity as developing countries bear the consequences of climate change resulting from developed countries' emissions. Second, it would open an opportunity for the United States to provide assistance for international humanitarian concerns as part of existing development goals. Third, international engagement could help to address national security issues that will arise from climate change. Fourth, coordination among countries could improve the effectiveness of adaptation efforts by reducing redundant activities or those that act at cross-purposes. Fifth, international engagement offers the United States opportunities to exchange lessons learned from the adaptation experiences. And sixth, international engagement would open expanded global market opportunities for U.S. adaptation technologies, systems, and services.

For these reasons, it is important to integrate climate change adaptation objectives into a range of foreign policy, development assistance, and capacity-building efforts. Overall, devising solutions and making decisions about adaptation options should be placed within a broad international context.

SCIENCE AND TECHNOLOGY ADVANCES NEEDED TO SUPPORT ADAPTATION CHOICES

America's climate choices in adapting to impacts of climate change are limited by the nation's insufficient knowledge of adaptation, tools, and options related specifically to climate change. The report suggests a broad agenda of science and technology needs. Examples range from a better understanding of how adaptation measures may interact with one another and contribute to overall goals for sustainability to research and development related to water use efficiency improvement. Significant improvements in capacities for adaptation analysis and assessment, adaptation option identification and development, and adaptation management and implementation are needed to broaden and strengthen our adaptation choices. Finally, to better manage and implement adaptation measures, it is important to improve risk-analysis techniques and observing systems that measure the magnitude of climate change and the effectiveness of adaptation actions.

As a component of a cross-agency climate change research program, the report suggests that climate change adaptation research and development should be pursued as a shared partnership among the federal government, other levels of government, the private sector and other NGOs, and the academic research community. Ideally, the program's scope would include studies of autonomous adaptation as well as planned adaptation; it should explicitly include monitoring and learning from ongoing experiences with adaptation in practice to build the knowledge base that can guide future adaptation planning and implementation; and it should expedite advances in adaptation science and technology that have promise in reducing critical national and regional vulnerabilities to climate change impacts in the coming decades.

CONCLUSIONS AND RECOMMENDATIONS

Because impacts of climate change are already being observed in the United States and elsewhere in the world, and because these impacts will increase in severity even if GHG emissions are reduced substantially in the near term, the United States needs to improve its ability to adapt to impacts of climate change. Concerns about these impacts are generating increasing interest in adaptation and wide-ranging discussions about potential actions that might be taken by individuals, sectors, cities, and states—in some cases without sufficient information about the options that are available.

It is the judgment of this panel that anticipatory climate change adaptation is a highly desirable risk-management strategy for the United States. Such a strategy offers potential to reduce costs of current and future climate change impacts, not only by realizing and supporting adaptation capacities across different levels of government, different sectors of the economy, and different populations and environments, but also by providing resources, coordination, and assistance in ensuring that a wide range of distributed actions are mutually supportive. Placed in a larger context of sustainable development, climate change adaptation can contribute to a coherent and efficient national response to climate change challenges that encourages linkages and partnerships across boundaries between different sectors and institutions in our society.

The report presents a number of findings and recommendations (see Box S.1) regarding the need for a national climate change adaptation effort. It emphasizes the term "national" rather than "federal" because adaptation is an inherently diverse and disaggregated process. Adaptation options themselves are immensely diverse, and choosing "how" and "when" to adapt from a long list of possible options requires careful evaluation of the socioeconomic context, the vulnerability of the sector or region, the resources available, and the scale at which the impact is likely to be felt. There is no one-size-fits-all adaptation option for a particular climate impact across the nation; instead, decision makers within each level of government, within each economic sector, and within civil society need to weigh the many tradeoffs between the available adaptation choices. Most decisions about how and when to implement adaptation options will require local input, and in many (if not most) cases, adaptation projects will occur at the local level. In addition, there is a very limited knowledge base evaluating adaptation measures. For all of these reasons, this report does not recommend specific adaptation measures to be implemented, aside from recommendations for several federal agencies. Rather, examples of adaptation measures that can be considered are discussed and a process for decision makers to develop and evaluate options for adapting is detailed.

The recommendations begin with a call for all decision makers—within national, state, tribal, and local agencies and institutions, in the private sector, and NGOs—to identify their vulnerabilities to climate change impacts and the short- and longer-term adaptation options that could increase their resilience to current and projected impacts. They call for the development of a collaborative national adaptation strategy and program, including a significant climate change research effort as part of an integrated climate change research initiative. They suggest adaptation planning and implementation by U.S. states and tribes, local governments, and the private sector, nongovernmental institutions, and society at large, in a spirit of national partnership; and they suggest U.S. support for international adaptation programs. Finally, they suggest incorporating adaptation objectives into a number of existing federal government programs.

In conclusion, the process of adapting to likely climate change impacts poses a daunting challenge and the stakes are high. Nevertheless, there are a large number of adaptation options that can be identified and initiated now. In many cases, these options would be relatively inexpensive, would be low-risk, would be consistent with sustainability principles, and would have multiple ancillary benefits. The recommendations listed in Box S.1 provide a solid framework within which the nation can initiate a national effort to adapt to the impacts of a changing climate. Along with initiating near-term adaptation measures, it is important to consider adaptation to climate change impacts as a process that will require sustained commitment and a durable yet flexible strategy for several decades to come.

BOX S.1
Recommendations

Recommendation 1: All decision makers—within national, state, tribal, and local agencies and institutions, in the private sector, and nongovernmental organizations (NGOs)—should identify their vulnerabilities to climate change impacts and the short- and longer-term adaptation options that could increase their resilience to current and projected impacts.

Recommendation 2: The executive branch of the federal government should initiate development of a collaborative national adaptation strategy, which might take the form of a national adaptation plan. The strategy (or plan) should be developed in partnership with congressional leaders, selected high-level representatives of relevant federal agencies, states, tribes, business and environmental organizations, and local governments and community leaders.

Recommendation 3: Federal, state, and local governments, together with nongovernmental partners, should work together to implement a national climate change adaptation program pursuant to the national climate adaptation strategy.

Recommendation 4: As part of an integrated climate change research initiative, the federal government should undertake a significant climate change adaptation research effort designed to provide a reliable foundation for adapting to the impacts of climate change in a larger context of sustainability.

Recommendation 5: Adaptation planning and implementation at the state and tribal level should be initiated regardless of whether the federal government provides the necessary leadership. States and tribes will need to take a significant leadership and coordination role, especially in areas where cities and other local interests have not yet established adaptation efforts. State and tribal governments should develop and implement climate change adaptation plans to guide policy and coordinate with federal, regional, local, and private-sector efforts pursuant to the national climate adaptation strategy.

Recommendation 6: Local governments should develop and implement climate change adaptation plans pursuant to the national climate adaptation strategy, in consultation with the broad range of stakeholders in their communities.

Recommendation 7: The private sector, NGOs, and society at large should assess their own vulnerabilities and risks due to climate change and actively engage and partner with the respective governmental adaptation planning efforts to help build the nation's adaptive capacity.

Recommendation 8: The United States should engage as a major player in adaptation activities at the global scale. The United States should support the establishment of a collaborative, sufficiently funded, international adaptation program that can be sustained over time.

Recommendation 9: Adaptation objectives should be incorporated into existing U.S. government programs and policies that have international components such as (1) agriculture, trade policy, and food security; (2) energy policy; (3) transportation policy; (4) international aid and disaster relief; (5) national security; and (6) intellectual property agreements for technology transfer to other countries.

Recommendation 10: Federal, state, and local entities and the private sector should take actions now to address current, known climate change impacts and risks and/or to provide effective risk management at a relatively low cost.

CHAPTER ONE

Introduction

Adaptation to climate change requires attention now because impacts are already being felt across the United States and further impacts are unavoidable, regardless of how immediately and stringently greenhouse gas (GHG) emissions are limited (IPCC, 2007a; USGCRP, 2009). Adaptation to climate *variability* is nothing new to humanity, but it now seems very likely that climate conditions by the later part of the 21st century will move outside the range of past human experiences (IPCC, 2007b; Solomon et al., 2009). Therefore, historical records and past experience are becoming incomplete guides for the future, and adaptation to climate change needs to become a high national priority. Either we adapt by mobilizing to reduce sensitivities to climate change and to increase coping capacities now, or we will adapt by accepting and living with impacts that are likely in many cases to disrupt our lives and livelihoods. The questions are how, where, and when to adapt—and whether in some cases, if climate change is relatively severe, we may face limits on our ability to avoid painful impacts by adapting.

ADAPTATION: KEY QUESTIONS, CHALLENGES, AND OPPORTUNITIES

Why Consider Adaptation Now?

Society and nature have always adjusted to climate *variability* and weather extremes, but climate *change* is moving climate conditions outside the range of past human experiences (IPCC, 2007b; Solomon et al., 2009). While previous experience in coping with climate variability or extremes can provide some valuable lessons for adapting to climate change, there are important differences between coping with variability and planning for climate change. Climate change has the potential to bring about abrupt changes that push the climate system across thresholds, creating novel conditions (Lenton et al., 2008). Likewise, thresholds in ecosystems (Adger et al., 2009; CCSP, 2009a) and human systems could be crossed, potentially overwhelming their adaptive capacity.

The prospect that the climate system, ecosystems, or human systems may experience significant transitions to new states renders our previous experience an incomplete guide for future adaptation. For example, it is unclear whether managing natural ecosystems for resilience (i.e., assisting ecosystems to return to a previous natural state after a disturbance) will remain a valid approach under future climate conditions, because ecosystems might cross thresholds into new steady states (West et al., 2009). Thus, managing certain ecosystems toward a new "natural" state might be a more viable strategy. Because of the potential for crossing such thresholds, adaptation to climate change begins with building adaptive capacities, frameworks, and institutional structures that can cope with future conditions that are beyond past experience.

Adaptation requires maintaining a long-term perspective because there are considerable uncertainties in estimating the nature, timing, and magnitude of climate impacts. This uncertainty involves the trajectory of future emissions and resulting changes in the mean climate conditions and the range of climate variability as well as other factors. Translating information at a global scale to local and regional scales can also contribute to uncertainty. Thus, precise predictions of many climate impacts are not available, despite the fact that the probability of some impacts is relatively high (e.g., loss of snowpack in the West and an ice-free Arctic in the summer). Adaptation to climate change calls for a new paradigm for considering a range of possible future climate conditions and associated changes in human systems and ecosystems, and for managing risks by recognizing prospects for departures from historical conditions, trends, and variation. This does not mean waiting until uncertainties have been reduced to consider adaptation actions, because there is a real risk that impacts could emerge too rapidly or too powerfully for delayed adaptations to reduce major disruptions to human and natural systems. Mobilizing now to increase the nation's adaptive capacity can be viewed as an insurance policy against an uncertain future. (See Box 1.1 for definitions of key terms used in this report.)

What Are the Risks?

Across the vast area of the United States and islands located within U.S. territory, many regions, sectors, populations, or resources exhibit vulnerabilities to climate variations and change (Figure 1.1). A recent report of the U.S. Global Change Research Program (USGCRP) highlights the range of climate change impacts on the United States (2009). Areas of particular concern include low-lying coastlines, especially coastal areas of the Southeast that are susceptible to hurricanes, sea level rise, saltwater intrusion, and land subsidence; the West, where water supplies are largely dependent on snowpack, particularly those with little storage relative to annual flow; inner cities in the Midwest and Northeast, where many residents do not have access to air conditioning; natural ecosystems and native villages in northern Alaska that are subject to rapid changes in temperature, thawing of permafrost, and loss of sea ice; and Western forests that are susceptible to wildfire and bark beetle infestation. In the absence of adaptation,

---

> ## BOX 1.1
> ## Definitions of Key Terms
>
> **Adapt, Adaptation:** Adjustment in natural or human systems to a new or changing environment that exploits beneficial opportunities or moderates negative effects.
>
> **Adaptive Capacity:** The ability of a system to adjust to climate change (including climate variability and extremes) to moderate potential damages, to take advantage of opportunities, or to cope with the consequences.
>
> **Resilience:** A capability to anticipate, prepare for, respond to, and recover from significant multi-hazard threats with minimum damage to social well-being, the economy, and the environment.
>
> **Risk:** A combination of the magnitude of the potential consequence(s) of climate change impact(s) and the likelihood that the consequence(s) will occur.
>
> **Vulnerability:** The degree to which a system is susceptible to, or unable to cope with, adverse effects of climate change, including climate variability and extremes. Vulnerability is a function of the character, magnitude, and rate of climate variation to which a system is exposed, its sensitivity, and its adaptive capacity.

FIGURE 1.1 An illustration of the range of climate change impacts across the United States. SOURCE: International Mapping Associates.

the risks of negative consequences that could accompany these types of impacts are heightened.

How Can We Adapt?

Because impacts of and vulnerabilities to climate change vary greatly across regions and sectors, adaptation decisions are fundamentally place-based, occurring at multiple scales, from that of the individual household or firm, to cities, regions, states, tribes, corporations, and economic sectors, to the level of the federal government and agencies within it that manage land and other resources. Considering the range, and in some instances the severity of climate change risks, it seems clear that capacities currently available for adaptation at the local and state levels are inadequate to address risks to health, well-being, property, and ecosystem services in many regions of the United States. While some localities and states have attempted to formulate adaptation strategies, they often lack the information, resources, and decision-making tools to pursue these plans. Meanwhile, there is a growing recognition that a new collaborative national effort is needed in support of adaptation across all scales, nationally and internationally (GAO, 2009a; NRC, 2009a,b).

Choices regarding how and when to adapt vary greatly. Adaptation could involve an immediate mobilization to reduce vulnerability to climate change and increase adaptive capacity. Or adaptation could take the form of accepting, responding to, and living with impacts that could in many cases be disruptive of our lives and livelihoods. Developing a framework for selecting among these types of adaptation options is critical given that, in different locations or for different sectors, the same strategy may produce very different results. This report identifies a number of choices that are available and outlines a method for selecting options based on a risk-management approach.

Adaptation to climate change can be categorized as "autonomous" or "planned." Autonomous adaptations are actions taken voluntarily by decision makers (such as farmers or city leaders) whose risk management is motivated by information, market signals, co-benefits, and other factors. Planned adaptations are interventions by governments to address needs judged unlikely to be met by autonomous actions—often adaptations larger in scale and/or resource requirements. The public sector plays important roles in both cases. Governments support autonomous adaptation by providing information, shaping market conditions through taxes and other policies (along with their own market decisions), and helping to enlarge portfolios of technologies and other alternatives for decentralized actions. Governments can also act more directly by developing plans and strategies, providing resources, and undertaking projects (such as infrastructure development).

What Are the Challenges and Opportunities?

Research on how to adapt to the many changes in the climate system has lagged behind efforts to identify policies to limit GHG emissions for many reasons, including the perception that efforts to adapt might reduce the commitment to limiting GHGs and result in more challenges in the long term. One consequence has been that adaptation actions have not been widely considered, and knowledge about climate change adaptation has been incompletely developed.

Reluctance to invest in adaptation research and actions is partly due to the fact that climate change is a slow-onset, multigenerational problem while decision makers tend to focus on short-term concerns and benefits. There is considerable empirical evidence suggesting that when individuals and businesses plan for the future, they do not fully weigh the long-term benefits of investing in loss-reduction measures, especially if there is only a small likelihood of reaping financial returns. The upfront costs of these protective measures loom disproportionately high relative to delayed expected benefits over time, especially given discount rates that are usually applied to most public and private investment. However, the tendency for people to focus on the short run and to ignore low-probability events below their threshold level of concern can have severe long-run consequences (Kunreuther and Michel-Kerjan, 2009).

Because climate conditions by the later part of the 21st century will likely be outside the range of past human experiences, it is difficult to make decisions now about the full range of anticipated climate change impacts in 2050 or 2100. Policy makers will need to select options that are flexible enough not to inadvertently preclude other options that may be needed or become available in the future (Adger et al., 2009). An additional challenge is that climate change impacts are rarely the key drivers of vulnerability. Other factors determining vulnerability include existing inequalities, demographics, land use and economic changes, dwindling nonrenewable resources, public health, and institutional and technological change (IPCC, 2007a). Developing proactive strategies and planning processes that consider multiple perspectives, multiple stressors, multiple time horizons related to intergenerational equity issues, and multiple competing interests is a complex challenge, calling for broad collaborations and partnerships.

Despite these and other challenges, adaptation activities can produce many benefits that support other social objectives, such as sustainable development, public health, economic competitiveness, national security, and international cooperation. Risk management for climate change impacts often helps to address other stresses on human and natural systems as well, and attention to climate change adaptation aims and strategies can be a catalyst for increased attention to relatively long-term sustainability objectives and choices.

SCOPE AND PURPOSE OF THE REPORT

This study and the overall America's Climate Choices suite of activities respond to a request by the U.S. Congress for the National Oceanic and Atmospheric Administration (NOAA) to execute an agreement with the National Academy of Sciences to establish a committee that will "investigate and study the serious and sweeping issues relating to global climate change and make recommendations regarding what steps must be taken and what strategies must be adopted in response to global climate change, including the science and technology challenges thereof." This panel was charged to describe, analyze, and assess actions and strategies to reduce vulnerability, increase adaptive capacity, improve resilience, and promote successful adaptation to climate change in different regions, sectors, systems, and populations across the nation. The panel's report draws on a wide range of sources and case studies to identify lessons learned from past experiences, promising current approaches, and potential new directions.

The challenges of this panel's assignment to "provide advice about what to do about adaptation" in the United States are numerous. The panel chose to provide a "menu" illustrating the long list of options available for consideration in adaptation planning for specific sectors. It uses the concept of *risk management,* broadly construed, to frame the process of planning and selecting approaches to climate change adaptation. The panel has outlined a number of principles for developing an adaptation strategy that addresses issues within the boundaries of the United States as well as in the international context. The decision to select a risk-management framework reflects an emerging scientific consensus that the United States will not be able to eliminate all risks associated with climate change; however, if the nation prioritizes activities well, it will be possible to minimize negative impacts and maximize the opportunities associated with climate change. The decision also reflects the panel's perspective that the needed actions involve limiting risks in the broadest sense—including risks associated with current and future political, economic, social, and environmental realities.

PRINCIPLES TO GUIDE CLIMATE CHANGE ADAPTATION

This report views and discusses adaptation through the lens of long-term sustainability and emphasizes cross-sectoral integration and an inclusive approach, because adaptation choices are linked directly to choices about limiting GHG and sustainable use of resources. It also recognizes the inevitability of tradeoffs and value judgments associated with all adaptation choices (Adger et al., 2009; Anderies et al., 2004; Tainter, 2003). For example, it is a fact that climate change "involves harm to some—now and in the future—on the basis of gain to others (in the past, present and future)" (Adger et al., 2009). Therefore, adaptation choices and decisions require both a scientific assessment of impacts and socioeconomic vulnerabilities and an assessment of their sensitivity to values and political decisions.

In preparing for its assignment, the panel recognized that its assessments and conclusions would be shaped by the values that its members brought to the group process. It chooses, therefore, to explicitly state the principles that guided the panel's work and to offer them as a possible set of criteria by which adaptation plans, policies, and adaptation options might be evaluated by others:

1. *In making adaptation decisions, focus not only on optimizing conditions for the current generation, but also look several generations ahead and consider ways to reduce risk over time.* Some adaptation decisions must be taken today, but planning needs to focus toward the future, when the risks from climate change will be greatest. It follows that we must guard against the possibility that current actions could actually exacerbate either exposure or sensitivity of future generations to these growing risks.
2. *Account for the impacts of adaptation decisions on natural and social systems as well as on individuals, firms, government institutions, and infrastructure.* For example, energy infrastructure, production processes, ecosystems, and emergency response capacity have complex and multiple interrelated components, yet their capacity to function needs to be protected and/or enhanced in the context of adaptation. The mechanisms established for ongoing evaluation of progress need to include assessments of effects on such synergistic systems.
3. *Recognize that ecosystem structure and functioning are particularly vulnerable to climate change and need consideration in adaptation decisions.* While human systems can use advanced technology and mobility to adapt, some ecosystem components are relatively limited in their ability to adapt to rapid rates of change. The intimate dependence of humans on the vital services provided by natural ecosystems needs to be recognized.

4. *Evaluate solutions from a perspective of sustainability so that social, economic, and environmental ramifications of proposed strategies and actions are explicitly recognized.* Adaptation decisions should be integrated into the broader context of sustainable development.
5. *Acknowledge equity and justice in adaptation decisions; there is a need to prioritize helping those with a higher degree of vulnerability to become more resilient.* The capacity to adapt is a critical feature of the nation's ability to respond to climate change. There are vulnerable populations and ecosystems in the nation, and their welfare is considered a high priority in adaptation actions. Likewise, while considering international investments, reducing risk and vulnerability in other countries is considered a high priority.
6. *There is a need to identify the potential impacts of proposed adaptation options on all affected parties.* It is important to consider the expected costs and benefits from adaptation programs to those who are affected by them, including the potential for unintended consequences.
7. *Develop a portfolio approach for addressing adaptation problems, including a suite of technology and social-behavioral-economic options.* The same underlying reality that speaks to the need for diversification in the financial sector applies to climate change response strategies.
8. *Develop methods of evaluation so that the risk of inactions can be compared with the risk of proposed actions.* The implications of proposed actions for public policy need to be recognized and explored so that decision makers can clearly see tradeoffs expressed not only in terms of costs and benefits but also in terms of short- and long-term risks.
9. *Recognize the international implications of U.S. adaptation and emissions-reduction efforts, as well as the impacts on the United States of decisions made by other countries.* The success of U.S. adaptation and mitigation efforts is in large part dependent on cooperative efforts across the globe.

ORGANIZATION OF THE REPORT

Identifying adaptation options and strategies to respond to climate change requires an understanding of anticipated changes in temperature and other climate variables and how these will in turn affect economic sectors and natural and human systems. In Chapter 2, the panel explores the impacts of projected changes in temperature and precipitation on natural resources, infrastructure, human health, and the environment. The chapter also identifies the scientific challenges that remain in assessing climate change impacts and vulnerabilities for adaptation.

Chapter 3 addresses the panel's charge to identify short-term options for adapting to climate change at different government levels and in different sectors by examining ongoing domestic and international adaptation activities for lessons learned. A menu of options for adapting to climate variability and other stressors is provided for particular decision needs in various sectors: ecosystems, agriculture and forestry, water, health, transportation, energy, and coastal regions. Furthermore, the chapter emphasizes the importance of comprehensive strategies to address multiple stresses, to increase the efficient use of adaptation resources, and to avoid inadvertent maladaptive actions.

In Chapter 4, the report highlights impediments to adaptation and an approach to overcoming these challenges. The approach involves an adaptive risk-management framework combining a portfolio of adaptation and emissions-reduction strategies, all of which should include provisions for learning by doing. The report draws on the example of New York City's adaptation efforts to illustrate how to develop an action-oriented adaptation strategy, principles for developing an adaptation plan, methods for selecting adaptation options, and tools and decision makers necessary for implementing an adaptation plan. Research needs to expand on adaptation opportunities are also identified.

Chapter 5 addresses the panel's charge to identify long-term strategies and opportunities to adapt to climate change. It demonstrates that effective adaptation will require the development of the capacity to adapt, which includes not only infrastructure and other investments but also more flexible institutions and investments in both adaptation processes and research on adaptation processes and outcomes. The chapter discusses the current lack of institutional capacity to build and implement an effective national adaptation effort, even to support the most vulnerable regions and sectors (those that are affected most immediately and severely by climate change). Because the nation lacks experience in and knowledge about how to adapt to rapid changes in the climate system, the chapter describes how adaptation capacity can be optimized through ongoing assessments of vulnerability and of the effectiveness of alternative adaptation options. Finally, Chapter 5 identifies the role of various decision makers at local, state, and federal government levels, as well as in the private sector, in building adaptive capacity and implementing climate change adaptation.

Chapter 6 focuses on the opportunities and rationale for considering adaptation within the international context. U.S. relationships with other countries will be affected in numerous ways by choices made in regards to addressing national and international resilience to climate change and supporting adaptation in especially vulnerable areas. At a fundamental level, the decisions made by individual governments are linked to impacts in other countries, through effects on the climate system, the global economy, and multiple other ways. The chapter concludes by highlighting the benefits of integrating climate change adaptation objectives into a range of foreign policy, development assistance, and capacity-building efforts, to improve the nation's ability to influence a broader range of outcomes, including economic and national security considerations.

Chapter 7 discusses the need for focused science and technology improvements to support adaptation activities, including evaluation of both gradual climate change and potential abrupt tipping points. It further elaborates on major challenges for adaptation research, including an improved understanding of human behavior affecting adaptation measures and how climate change adaptation relates to questions of sustainability. The chapter identifies a number of sector-specific adaptation options that would benefit from science and technology advances.

Chapter 8 provides a summary of the panel's conclusions and recommendations. The chapter emphasizes a need for broad-based national collaboration in planning and implementing adaptation actions, and it examines opportunities for near-term actions that would enhance the nation's adaptive capacity. This has profound meaning for both the near term and the long term. In the near term, America's choices will be focused mainly on adaptation actions that reduce risks from climate change impacts while at the same time helping to meet other needs, such as reducing risks from climate variability or reducing threats that could undermine near-term economic and social development. The emphasis will be on how climate change adaptation offers co-benefits as it reduces vulnerabilities related to ecosystem stress, water resource management, community resilience, human health, energy security, and other social concerns. This report identifies a wide variety of possible adaptation actions, some of which represent relatively low net costs to decision makers in many locations and sectors and have the potential for co-benefits (Chapter 3), along with strategies for identifying and assessing such possible actions (Chapter 4) and opportunities for institutional partnerships as the nation seeks to work together in its response to climate change impacts (Chapter 5).

In the longer term, America's choices will be focused mainly on three needs: (1) ensuring the development of adaptive institutions that continually consider further actions to cope with longer-term impacts and vulnerabilities, and also ensuring that these institutions are supported by systems that monitor emerging climate conditions and their effects and provide feedback about experiences with climate change adaptation (Chapters 4 and 5); (2) enlarging our range of choices by strengthening the science and technology that open new options for action and significantly improve our knowledge of their benefits, costs, potentials, and possible limits (Chapter 7); and (3) sharing the responsibility for supporting adaptation to climate change impacts in other areas of the world that are not capable of adapting on their own (Chapter 6). Through the pursuit of these near- and long-term adaptations, America can minimize harm and take advantage of opportunities that may result from a changing environment while sustaining human welfare and ecological integrity.

CHAPTER TWO

Vulnerabilities and Impacts

A daptation is intended to reduce climate change vulnerabilities and impacts. That means any consideration of adaptation planning must begin with consideration of risks associated with climate change vulnerabilities and impacts, to the extent that these can be anticipated.

More specifically, adaptation includes (1) the strategies, policies, and measures implemented to avoid, prepare for, and effectively respond to the adverse impacts of climate change on natural and human systems (to the extent that they can be anticipated), and (2) the social, cultural, economic, geographic, ecological, and other factors that determine the vulnerability of places, systems, and populations. Climate-related changes can create new or interact with existing vulnerabilities to cause impacts, including changes in:

- Temperature, both averages and extremes;
- Precipitation, both averages and extremes;
- The intensity, frequency, duration, and/or location of extreme weather events;
- Sea level; and
- Atmospheric carbon dioxide (CO_2) concentrations.

Vulnerability is often defined as the capacity to be harmed. It is a function of the character, magnitude, and rate of climate variation to which a system is exposed, its sensitivity, and its adaptive capacity (Clark et al., 2000; IPCC, 2007a; Turner et al., 2003). Vulnerabilities can be reduced by limiting the magnitude of climate change through actions to limit greenhouse gas (GHG) emissions (*ACC: Limiting the Magnitude of Future Climate Change*; NRC, 2010c), reducing sensitivity (the underlying social, cultural, economic, geographic, ecological, and other factors that interact with exposures to determine the magnitude and extent of impacts), or improving coping capacity (the ability to avoid, prepare for, and respond to an impact so that it is not seriously disruptive). Actions to reduce sensitivity and increase coping capacity are keys to effective adaptation to climate change.

A risk perspective (Chapter 4) considers the probability of an exposure and its consequences, including uncertainties in projecting the magnitude, rate, and extent of climate change. It also considers factors that shape sensitivities and coping capacities, which are as important as exposures in determining impacts. Later chapters of this report consider options for reducing risks by reducing sensitivities and improving coping capacities. This chapter provides the context by summarizing what is known about current and projected climate change impacts and vulnerabilities in the United States.

PROJECTED U.S. CLIMATE CHANGES THAT COULD REQUIRE ADAPTIVE RESPONSES

Climate-related impacts that require adaptation are already being observed in the United States and its coastal waters (USGCRP, 2009), and empirical evidence suggests that many of these and other impacts will grow in severity in the future (USGCRP, 2009). Over the past 50 years:

- Average temperature in the United States increased more than 2°F (1°C).
- Precipitation in the United States increased an average of 5 percent, and the intensity of precipitation events also increased.
- Many types of extreme weather events increased in frequency and intensity; hurricanes, although not more frequent, increased in destructive energy.
- Sea level increased along most of the U.S. coast over the past 50 years, with some areas along the Atlantic and Gulf coasts experiencing increases of greater than 8 inches.
- Arctic sea ice extent decreased 3 to 4 percent per decade, with end-of-summer ice declining at 11 percent per decade.

These changes are causing impacts that should promote adaptation regardless of whether the trends are permanent. In many circumstances, projected increases in the frequency and intensity of many extreme weather events over the next several decades will initially drive adaptation more than changes in mean weather variables. Images from Alaska provide a vivid example of observed climate change impacts in the United States (Figure 2.1). These impacts already require adaptation in many locations and economic sectors.

Effective adaptation depends on understanding projected climatic changes at geographic and temporal scales appropriate for the needed response. The report *Global Climate Change Impacts in the United States* (USGCRP, 2009) was based on two projected climate change scenarios: one of relatively moderate changes in the event that GHG emissions peak before the middle of the century and decline thereafter (lower emissions scenario), and another of relatively severe changes in the event that GHG emissions continue to grow at current rates without aggressive actions to limit them (higher emissions scenario). Prospects for adaptation to keep disruption from climate change impacts at socially acceptable levels depend very substantially on what happens with efforts to limit emissions. At moderate rates and levels of climate change, adaptation can be very effective. At severe rates and levels of climate change, limits of

FIGURE 2.1 The Arctic village of Shismaref: Rising sea levels and fierce storms have eroded the shoreline near this coastal Inupiat village, breaking down sea walls and washing away homes. Residents decided to relocate farther inland for safety, giving up their traditional fishing, sealing, and home-building sites. SOURCE: Photo by Edward W. Lempinen/AAAS. © 2006 AAAS.

many adaptation options are likely to be reached, and resulting adaptations are likely to be much more disruptive.

A key fact about climate change impacts is that stabilization of atmospheric GHG concentrations will not immediately stabilize the climate, which will continue to change for some time because of the delayed response of the climate to the buildup of GHGs emitted in the recent past. A companion to this report (*ACC: Limiting the Magnitude of Future Climate Change*; NRC, 2010c) details the challenges and choices the nation and the world face in sufficiently limiting GHG emissions to keep climatic changes at a relatively moderate level.[1] It also concludes that stabilizing emissions at moderate levels is becoming increasingly difficult in the face of U.S. and global inaction. The U.S. Global Change Research Program's (USGCRP's) characterizations of two possible futures (lower or higher emissions) show that an effective response to climate change must include both adaptation and mitigation (e.g., Wilbanks and Sathaye, 2007).

Much of the current knowledge about projected climate changes in the United States comes from an assessment process mandated by the U.S. Congress in the Global

---

[1] For a discussion linking emission rates and atmospheric concentrations of GHGs to changes in global mean temperature, see Chapter 2 of *ACC: Limiting the Magnitude of Future Climate Change* (NRC, 2010c).

Change Research Act of 1990 (P.L. 101-606), including the U.S. National Assessment (USGCRP, 2001) and conclusions from 21 widely peer-reviewed Synthesis and Assessment Products (SAPs) produced by the U.S. Climate Change Science Program between 2006 and 2009 on specific topics ranging from knowledge of the physical climate system to the interface between climate change and society. The SAPs were summarized and updated in *Global Climate Change Impacts in the United States* (USGCRP, 2009). According to this summary report, future climate change impacts in the United States will include warmer average temperatures, changes in precipitation patterns, more frequent heat waves and severe storms, rising sea level, and decreases in sea ice and permafrost, which will be particularly rapid in the Arctic.

The average temperature in the United States will continue to rise with climate change, but the magnitude of the increase depends primarily on the amount of heat-trapping GHGs emitted globally and how sensitive the climate is in responding to those emissions. Figure 2.2 shows projected temperature change under the higher and lower emissions scenarios in midcentury and at the end of the century. The brackets on the thermometers represent the likely range of model projections, although lower or higher outcomes are possible. By the end of the century, the average U.S. temperature is projected to increase approximately 7°F to 11°F (4°C to 6°C) under the higher emissions scenario and approximately 4°F to 6.5°F (2°C to 4°C) under the lower emissions scenario (USGCRP, 2009).

Projections of future precipitation generally indicate that northern areas (higher latitudes) will receive more precipitation, and southern areas, particularly in the West, will become drier (USGCRP, 2009). However, the mechanisms by which human-induced climate change affects precipitation are subtler than those of temperature and paint a more complex picture (e.g., Zhang et al., 2007). Figure 2.3 shows projected changes by 2080-2099 under the higher emissions scenario; these are sample results from climate models, not projections of certainty for the future.

The amount of rain falling in the heaviest downpours has already increased approximately 20 percent on average in the past century, and this trend is very likely to continue, with the largest increases in the wettest places (USGCRP, 2009). Figure 2.4 shows projected changes from the 1990s average to the 2090s average in the amount of precipitation falling in light, moderate, and heavy events. The lightest precipitation is projected to decrease, while the heaviest will increase, continuing the observed trend.

Many types of extreme weather events, such as heat waves, have become more frequent and intense during the past 40 to 50 years, while cold extremes have become less frequent (USGCRP, 2009). In the future, currently rare extreme events (for example a 1-in-20-year event) are projected to become more commonplace (Figure 2.5), al-

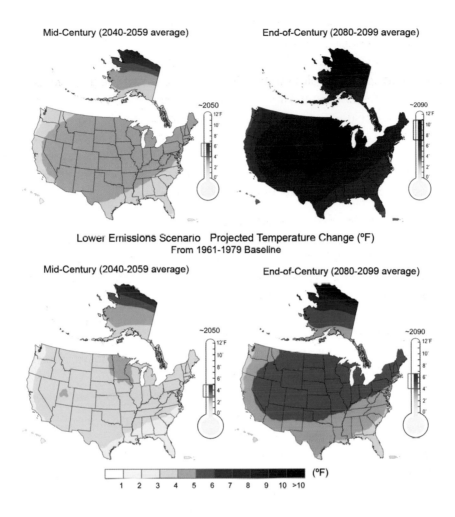

FIGURE 2.2 Projected temperature change (°F) from 1961-1979 baseline. NOTE: These results are derived from global models whose spatial resolution is insufficient to resolve important details like mountain ranges. SOURCE: USGCRP (2009) (*http://www.globalchange.gov*).

though these projected increases will not be uniformly distributed over temporal and spatial scales. For example, a day so hot that it is currently experienced once every 20 years would likely occur every other year or more frequently by the end of the century under the higher emissions scenario. Although uncertainties remain about whether the number of hurricanes could increase with climate change, the destructive energy of Atlantic hurricanes is likely to increase in this century as sea surface temperature rises (USGCRP, 2009) (Figure 2.6). In addition, cold-season storm tracks are shifting

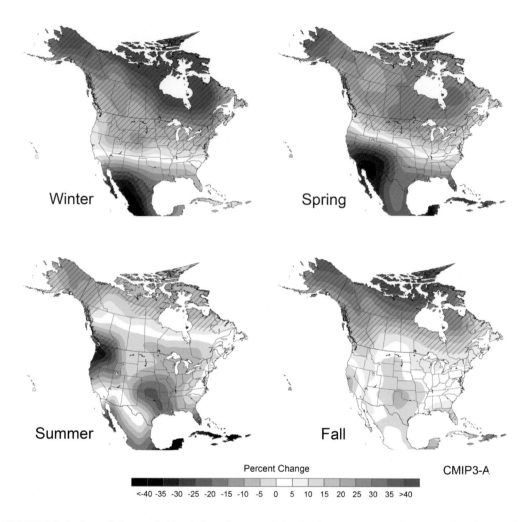

FIGURE 2.3 Projected change in North American precipitation by 2080-2099. NOTE: Cross-hatching indicates areas in which climate models do not agree. SOURCE: USGCRP (2009) (*http://www.globalchange.gov*).

northward, and the strongest storms are likely to become stronger and more frequent (USGCRP, 2009).

The ocean is warming and glaciers and polar ice sheets are melting, causing sea level to continue to rise, most likely at a faster rate than in recent history. Globally, under the higher emissions scenarios, average sea level is estimated to rise by 3 to 4 feet (USGCRP, 2009). How much land will become submerged will vary regionally, depending on the regional tectonics and geomorphology (land masses can be in the process

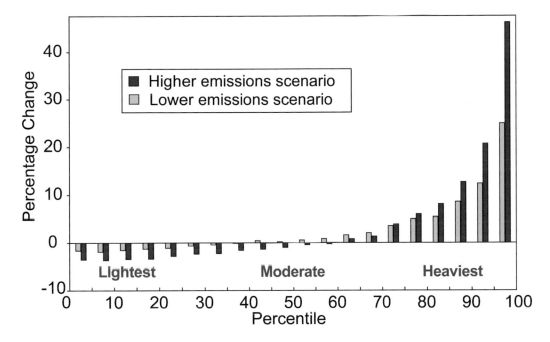

FIGURE 2.4 Projected changes in light, moderate, and heavy precipitation from the 1990s average to the 2090s average in North America. As shown here, the lightest precipitation is projected to decrease, while the heaviest will increase, continuing the observed trend. The higher emissions scenario yields larger changes. Projections are based on the models used in the IPCC (2007) Synthesis Report. NOTE: "Lower emissions scenario" refers to IPCC SRES B1, "higher emissions scenario" refers to A2, and "even higher emissions scenario" refers to A1FI. SOURCE: USGCRP (2009) (*http://www.globalchange.gov*).

of rising or sinking relative to sea level) and ocean currents (which can cause the ocean surface to rise or sink relative to the average global sea level).

DETERMINING VULNERABILITIES TO PROJECTED CLIMATE CHANGES

As defined earlier, vulnerability is a function of the character, magnitude, and rate of climate change to which a system is exposed, as well as the system's sensitivity and its adaptive capacity. Therefore, vulnerability can be assessed through the examination of these three factors. Assessing exposure to climate change reveals regional differences in the climate-related impacts that the United States will experience. Table 2.1 summarizes climate-related exposures and the regions that will most likely be affected.

Vulnerability can also be examined through the sensitivity and adaptive capacities of

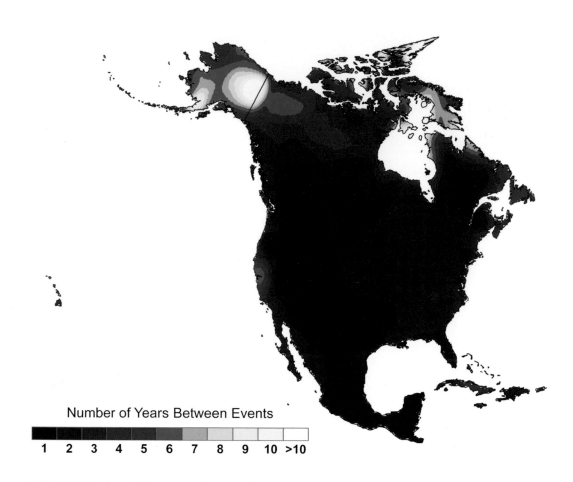

Number of Years Between Events

| 1 | 2 | 3 | 4 | 5 | 6 | 7 | 8 | 9 | 10 | >10 |

FIGURE 2.5 Projected frequency of extreme heat (2080-2099 average). SOURCE: USGCRP (2009) (*http://www.globalchange.gov*).

a particular community, system (i.e., economic, ecosystem, etc.), or sector. Vulnerability encompasses the risk and protective factors that ultimately determine whether a subpopulation experiences adverse outcomes due to climate change (Balbus and Malina, 2009). For example, Table 2.2 summarizes various subpopulations that are particularly vulnerable to multiple climate-related exposures to health risks.

Vulnerabilities of human systems are shaped by a wide variety of nonclimatic conditions. Although important, limited access to financial resources is not the only source of vulnerability. Examples of other sources include population shifts and development choices, such as dense urban development in drought-prone areas; and places and

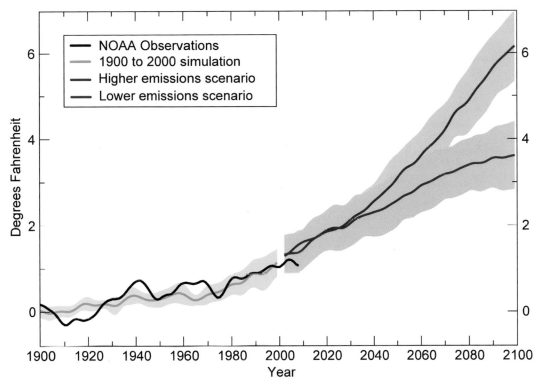

Observed (black) and projected temperatures (blue = lower scenario; red = higher scenario) in the Atlantic hurricane formation region. Increased intensity of hurricanes is linked to rising sea surface temperatures in the region of the ocean where hurricanes form. The shaded areas show the likely ranges while the lines show the central projections from a set of climate models.

FIGURE 2.6 Projected sea surface temperature change. SOURCE: USGCRP (2009) (*http://www.global-change.gov*).

communities that are especially dependent on climate-sensitive industries such as agriculture, forestry, and tourism. In the built environment, each city's residents and infrastructures will be affected in unique ways (USGCRP, 2009). The vulnerability of natural systems, on the other hand, depends primarily on an ecosystem's resilience to change. Changes in ecosystem function, in turn, affect human communities that depend on natural ecosystems to maintain clean water supplies, soil fertility, and other vital services (USGCRP, 2009).

TABLE 2.1 Summary of regional climate-related impacts

Climate-Related Impacts

United States Census Regions	Early Snowmelt	Degraded Air Quality	Urban Heat Island	Wildfires	Heat Waves	Drought	Tropical Storms	Extreme Rainfall with Flooding	Sea Level Rise
New England ME VT NH MA RI CT	●	●	●		●	●		●	●
Middle Atlantic NY PA NJ DE MD	●	●	●		●	●	●	●	●
East North Central WI MI IL IN OH	●	●	●		●	●		●	
West North Central ND MN SD IA NE KS MO	●		●		●	●		●	
South Atlantic WV VA NC SC GA FL DC		●	●	●	●	●	●	●	●
East South Central KY TN MS AL			●	●	●	●	●		●
West South Central TX OK AR LA		●	●	●	●	●	●	●	●
Mountain MT ID WY NV UT CO AZ NM	●	●	●	●	●	●		●	
Pacific AK CA WA OR HI	●	●	●	●	●	●	●	●	●

SOURCE: Adapted from CCSP (2008f).

TABLE 2.2 Summary of vulnerability to climate-sensitive health outcomes by subpopulation

Groups with Increased Vulnerability	Climate-Related Exposures
Infants and children	Heat stress, ozone air pollution, water- and food-borne illnesses, Lyme disease, dengue
Pregnant women	Heat stress, extreme weather events, water- and food-borne illnesses
Elderly/Chronic medical conditions	Heat stress, air pollution, extreme weather events, water- and food-borne illnesses, dengue
Impoverished/Low socioeconomic status	Heat stress, extreme weather events, air pollution, vector-borne infectious diseases
Outdoor workers	Heat stress, ozone air pollution, Lyme disease, other vector-borne infectious diseases

HOW CHANGING CLIMATE CONDITIONS AND VULNERABILITIES IMPACT DIFFERENT U.S. SECTORS

Climate Change Will Interact with Many Social and Environmental Stresses

Society, its infrastructure, and its policies were developed in a relatively stable climate. Although climate change will create advantages for some locations and populations, on average, climate change is expected to adversely affect water resources, ecosystems, human health, energy, transportation, and other sectors. The expectation of adverse impacts stems in part from the fact that these systems were designed during a period of relatively stable climate conditions and in part from the accelerating rate of change, which presents a novel challenge for adaptation. Recent events suggest that changes in extreme weather events, including heat waves, floods, droughts, windstorms, and wildfires, will likely be particularly challenging for communities and sectors to adapt to. The combination of climate change and trends in population growth also poses serious adaptation challenges. For example, population growth in the past century has been greatest in the South, along the coast, and in larger cities; this trend aligns somewhat with places where the threats of future heat waves and severe storms are greatest (USGCRP, 2009). With most of the U.S. population residing in urban areas, vulnerabilities associated with aging urban infrastructure, traffic congestion, air quality, social inequities, and other variables exacerbate the challenges of adapting to climate change. Key conclusions about how climate change will interact with social and environmental vulnerabilities are summarized by sector in the following sections that were derived from the SAPs and the USGCRP (2009).

Climate Change Will Place Additional Burdens on Already Stressed Water Resources

Climate change has already altered and will continue to alter the water cycle, affecting where, when, and how much water is available for various uses. Rising temperatures, for example, interact with other components of the climate system to alter patterns of precipitation. More intense droughts and flooding events are projected to become common in some regions. Increased droughts will have direct impacts on water resources, agriculture, and ecosystems, as well as leading to an increased risk of wildfires. Changes in precipitation will also alter runoff patterns (USGCRP, 2009) (Figure 2.7).

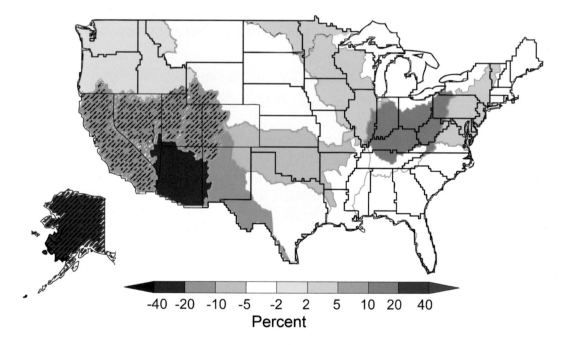

FIGURE 2.7 Projected changes in annual average runoff for 2041-2060 relative to a 1901-1970 baseline by water resource region, based on analyses using emissions that fall between the lower and higher emissions scenarios. Lower average runoff is expected in the Southwest and greater runoff is projected for the Northeast. Colors indicate percentage changes in runoff, with hatched areas indicating greater confidence due to strong agreement among model projections. SOURCE: USGCRP (2009) (*http://www. globalchange.gov*).

A greater challenge for much of the western water sector will be a decrease in total snowpack and an altered timing of seasonal flow in snowmelt-dominated river basins. Warmer temperatures already have increased the proportion of precipitation that falls as rain rather than snow in the West, resulting in less snowpack accumulation and earlier snowmelt. In the latter half of the 20th century, peak flows in western streams arrived 1 to 2 weeks earlier. By the late 21st century under the higher emissions scenario, peak flows are projected to arrive 2 to 5 weeks earlier than in 1951-1980, leading to lower summer stream flows, generally less water availability, and changes in surface and groundwater quantities (USGCRP, 2009).

These changes in precipitation, evaporation, and snowpack will further stress water resource allocations in regions such as the West and Southwest. For example, the Colorado River already has insufficient flow to support demand. Even under the lower emissions scenario, large areas of the Southwest are projected to receive 15 to 25 percent less spring precipitation by the end of this century. Under the higher emissions scenario, widespread decreases of 30 percent and more are projected. Greater conflicts over water resource allocations between agriculture, urban areas, and natural ecosystems are likely in many areas. Table 2.3 illustrates the varied impacts possible with changing water resources.

As discussed in later chapters, current resource management plans are based on historical climatic averages (e.g., stream flow, reservoir size, runoff) that will not continue

TABLE 2.3 Highlights of water-related impacts by sector

Sector	Examples of Impacts
Human health	Heavy downpours increase incidence of waterborne diseases and floods, resulting in potential hazards to human life and health.
Energy supply and use	Hydropower production is reduced due to low flows in some regions. Power generation is reduced in fossil fuel and nuclear plants due to increased water temperatures and reduced cooling water availability.
Transportation	Floods and droughts disrupt transportation. Heavy downpours affect harbor infrastructure and inland waterways. Declining Great Lakes levels reduce freight capacity.
Agriculture and forests	Intense precipitation can delay spring planting and damage crops. Earlier spring snowmelt leads to increased extent of forest fires.
Ecosystems	Cold-water fish are threatened by rising water temperatures. Some warm-water fish will expand ranges.

SOURCE: USGCRP (2009) (*http://www.globalchange.gov*).

under future climate conditions (Milly et al., 2008). In addition, societal vulnerability to future water stress and related conflicts is increased by some current laws and practices such as those governing interstate water allocation and reservoir operations, and by the relatively low price that is paid for water in most regions of the United States.

Climate Change Can Lead to Large Ecosystem Changes When Impact Thresholds Are Crossed

Changing climate conditions will change the distribution and migration patterns of plant and animal species, species productivity and abundance, and species interactions and habitat utilization. For example, wildlife corridors for migrations may shift. Species are already moving pole-ward and to higher elevations to remain within their optimal temperature ranges and, in the process, are invading new habitats. Predator-prey relationships are likewise being altered. Some changes will become irreversible as they cross certain threshold levels (i.e., species extinction; see Box 2.1) (USGCRP, 2009).

BOX 2.1
The Value of Biodiversity

Humans depend on biodiversity—the array of plants, animals, fungi, and microorganisms that constitute the fabric of the living world. All of our food comes directly or indirectly from plants and animals. Most people in the world depend on living organisms for their medicines; for those who obtain their drugs from pharmacies, roughly half are based on molecules found first in living organisms. Building materials, fossil fuels, chemical feedstocks, and future products and cures yet to be discovered—all of these are or will be derived from living organisms (Diaz et al., 2006). The communities and ecosystems that these organisms constitute support our lives through a variety of *ecosystem services*, activities that collectively determine the qualities of the atmosphere, regulate local climates, conserve topsoil, regulate runoff, provide pollination services for cultivated and wild plants, and contribute to the beauty and healthfulness of our lives (Daily, 1997).

Combined with human population growth and land use change, climate change is a direct threat to the diversity of plant and animal species in many parts of the world, forcing already stressed species to respond to changes in climatic conditions that exceed the rate of change experienced in the past. The value of biodiversity has been recognized by policy actions such as passage of the Endangered Species Act and creation of national parks and biosphere preserves. Climate change could make it difficult to preserve valued landscapes and many of the species that make them special.

Of the global ecosystem services assessed by the Millennium Ecosystem Assessment (MEA), 60 percent were already on the decline due to human-driven stresses on natural systems (MEA, 2005). Because of such stresses—including exploitation, contamination, and habitat fragmentation—many ecosystems on land and sea are thought to be less resilient to the additional stress of a changing climate (USGCRP, 2009). Thus, climatic changes might result in further declines in provisioning services (e.g., food or timber production), regulating services (e.g., shoreline protection from storms provided by wetlands), supporting services (e.g., water filtration and contaminant removal), and cultural services (e.g., recreation or sacred places). Loss or changes in these ecosystem services would negatively affect human well-being.

Many terrestrial and marine ecosystems are changing in character and may look fundamentally different in the future, with unknown consequences. Some ecosystems, such as shallow-water coral reefs, are at risk of disappearing completely in coming decades; other ecosystems are simply changing in unpredictable ways. Significant effects of climate change have been observed in ecosystem processes, such as those that control plant growth and decomposition. These and other changes could cause large-scale shifts in the ranges of species, the timing of the seasons, and animal migration (NRC, 2008a). Because species display great variation in their sensitivity to climate change, mobility, and lifespan, not all members of an ecological community will shift uniformly in response to climate change (USGCRP, 2009). The combination of climate change with other environmental stressors such as human resource exploitation (e.g., fishing or timber harvest) and barriers to migration (e.g., roads, residential developments, and other built environments) will increase the risk of species extinctions.

Changes in ecosystems are very likely to continue throughout the century. For example, Figure 2.8 shows current and projected shifts in forest types, with major changes projected for many regions. In the Northeast, under a midrange warming scenario, the currently dominant maple-beech-birch forest type is projected to be completely displaced by other forest types.

Wildfires, outbreaks of insect pests and disease pathogens, and spread of invasive weed species have already increased, with climate implicated in some of these changes. These trends are likely to continue. In the western United States, the frequency of large wildfires and the length of the fire season increased substantially in recent decades, due primarily to earlier spring snowmelt and higher spring and summer temperatures (although annual area burned may have been just as high in the 1920s). Insect pests, coupled with pathogens, annually cause an estimated $1.5 billion in damage in the United States. Changes in climate contributed significantly to several

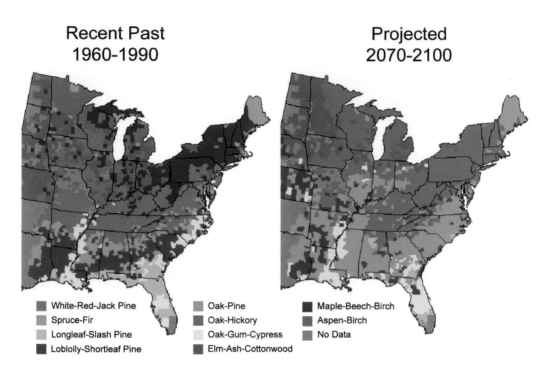

Recent Past
1960-1990

Projected
2070-2100

White-Red-Jack Pine
Spruce-Fir
Longleaf-Slash Pine
Loblolly-Shortleaf Pine

Oak-Pine
Oak-Hickory
Oak-Gum-Cypress
Elm-Ash-Cottonwood

Maple-Beech-Birch
Aspen-Birch
No Data

FIGURE 2.8 Current and projected shifts in forest types. SOURCE: USGCRP (2009) (*http://www.global-change.gov*).

major insect pest outbreaks in the United States and Canada over the past several decades, particularly the mountain pine beetle.

Deserts and drylands are likely to become hotter and drier, feeding a self-reinforcing cycle of invasive plants, fire, and erosion. The arid Southwest is projected to become even drier in this century, and emerging evidence suggests that this process is already under way. Deserts are also projected to expand to the north, east, and upward in elevation in response to projected warming and associated changes in climate.

Arctic marine ecosystems are being severely affected by the loss of summer sea ice, and further changes are expected. The ice currently provides a vital platform for ice-dependent seals (such as the ringed seal), polar bears, and walruses to hunt and rest. It is uncertain how these animals could adapt to significantly less sea ice.

As in other sectors, current policies to manage the impacts of human activities on the natural environment—including fisheries, wildlife, and forest management; pollution control; and habitat protection—were based on the assumption that the Earth's climate was relatively stable. The current rate of climate change suggests that management systems are likely to require substantial revision to maintain current levels of effectiveness in a warmer world (West et al., 2009).

Coastal Areas Are at Increasing Risk from Sea Level Rise and Storm Surges

The combination of sea level rise and storm surges poses a threat to coastal cities and ecosystems, especially areas that already experience multiple other stressors such as urban growth, human-induced changes in sediment loading and land subsidence, and high nutrient runoff. Uncertainties as to the exact extent of sea level rise in a particular location mean that local, state, and national agencies and organizations involved in coastal zone planning and management need to prepare for a range of possibilities.

Coastal counties are among the most densely populated areas in the United States— more than a third of all Americans live near the coast, and activities along or on the ocean contribute more than $1 trillion to the nation's economy. This intense development of coastal areas has increased their vulnerability to sea level rise and storm surges by decreasing the extent of natural buffers and causing accelerating rates of subsidence. For example, coastal Louisiana has already lost 1,900 square miles of wetlands in recent decades due to sea level rise and human alterations, weakening its capacity to absorb storm surges from hurricanes. Shoreline retreat has been observed along most U.S. exposed shores. Projected sea level rise could inundate portions of major cities such as Miami or New York during storm surges or even extreme high tides. Sea level rise can also lead to saltwater intrusion of freshwater aquifers in coastal areas that could reduce freshwater supplies (USGCRP, 2009).

Projected changes in the timing of spring runoff and associated high nitrogen loading from agriculture will combine with increased sea surface temperatures to further reduce available oxygen in coastal waters. An additional threat to marine ecosystem health is ocean acidification (decrease in ocean pH), which is caused directly by rising atmospheric CO_2. Already heat-stressed corals will be further damaged by ocean acidification, with impacts that might reverberate across the reef food web (see NRC, 2010d). In addition, island states, territories, and protectorates vulnerable to sea level rise face additional threats due to the potential loss of coral reefs that serve as natural buffers from storm surge.

Crop and Livestock Production Will Be Increasingly Challenged

Agriculture is considered one of the sectors most adaptable to climate change. In the United States, agricultural products contribute more than $200 billion in food commodities, with livestock accounting for more than half. Plants and animals display a broad range of vulnerability to increased temperature, resulting in diverse impacts. Global market pressures will also determine the ability of the agricultural sector to adapt to climate change.

At average global temperature increases of less than 5.4°F (3°C), some agricultural systems will benefit and others will be adversely affected. At higher levels of warming, crop and livestock production is projected to decline in all regions due to increased heat, pests and pathogens, water stress, and weather extremes (Rosenzweig et al., 2008). Many crops grow better at higher atmospheric CO_2 levels, but protein and nitrogen content often decline, resulting in less nutritious crops. Higher CO_2 levels also improve water-use efficiency in some plants, which could benefit agriculture in water-stressed areas. However, heat-related stresses will increase water demand and require adjustments in agriculture practices. In addition, livestock productivity rates are projected to decrease because many warm-blooded species, including milk cows, are susceptible to heat stress, and the quality of pasture and rangeland forage will decrease with higher CO_2 levels.

Threats to Human Health Will Increase

Climate change directly affects human physical and mental health through changes in the frequency, intensity, and/or duration of extreme weather events. While the frequency of extreme cold events will likely decrease, heat waves are increasing. Depending on the extent and effectiveness of adaptation measures, heat-related illnesses and deaths could increase over coming decades as heat waves increase in frequency, intensity, and duration (Kovats and Hajat, 2008) (see Box 2.2). Heat waves are already one of the leading causes of weather-related mortality, as evidenced in Europe during the summer of 2003 when extreme heat was linked to more than 70,000 excess deaths (Robine et al., 2008) and the 1995 Chicago heat wave that caused 696 excess deaths (Whitman et al., 1997; Semenza et al., 1999).

Although data are limited to estimate the health impacts that may result from changes in extreme weather events other than heat waves, such events create potentially serious health consequences. For example, flooding not only causes direct injuries but, in some regions, also increases the risk of sewage overflows that contaminate drinking water. Urban population growth is expected to exacerbate the health risks associated with extreme events.

Warmer air temperatures are associated with higher ozone levels, a known lung irritant. Because half of the U.S. population is already living in counties where air pollu-

---

> **BOX 2.2**
> **Urban Heat Waves**
>
> Throughout much of the Midwest, projections for 2090 suggest increases in nighttime temperatures (relative to 1975) of more than 3.6°F (2°C) during the worst heat waves (Ebi and Meehl, 2007). Illnesses caused by exposure to high temperatures include heat cramps, heat-induced fainting, heat exhaustion, heatstroke, and death (Kilbourne, 1997). Heatstroke has a high fatality rate, and even nonfatal heatstroke can lead to long-term illness (Dematte et al., 1998). Although the risk of heat illness exists for the entire population, a number of factors increase the risk: older and younger ages; use of certain drugs; dehydration; low level of fitness; excessive exertion; overweight; lower socioeconomic status; and living alone.
>
> A heat wave of the same magnitude as the 2003 European heat wave in a large American city is projected to increase excess heat-related deaths by more than five times the average (Kalkstein et al., 2008). New York City's total projected excess deaths would exceed the current national summer average for heat-related mortality, with the death rate approaching annual mortality rates for common causes of death such as accidents. The extent to which death rates would actually increase during an event will depend on adaptation, including the population's acclimatization to higher temperatures, modifications to the urban environment that reduce urban heat island effects, implementation of heat wave early warning systems, greater access to air conditioning, better education about response options, and other measures.

tion exceeds national health standards, further deterioration in air quality is a concern. There is growing evidence that ground-level ozone concentrations would be more likely to increase than decrease in the United States as a result of climate change, if one assumes that emissions of ozone precursors remain constant (Bell et al., 2007). An increase in ozone could cause or exacerbate heart and lung diseases. Warmer temperatures and higher CO_2 levels also are likely to increase pollen production and lengthen the pollen season for some plants, potentially affecting allergies and respiratory health (Beggs, 2004; Kinney, 2008; USGCRP, 2009).

The number of cases of climate-sensitive food- and water-borne diseases (i.e., salmonella) may increase among susceptible populations (USGCRP, 2009). The very young and old, the poor, those with health problems and disabilities, and certain occupational groups are at greater risk. Vector-borne diseases (i.e., Lyme disease and others) may shift their geographic ranges, although climate will seldom be the only factor (USGCRP, 2009).

Energy and Transportation Will Be Affected

Climate change impacts on the energy industry are likely to be most apparent at subnational scales, such as regional effects of extreme weather events, reduced water availability leading to constraints on energy production, and sea level rise affecting energy production and delivery systems. Warming will be accompanied by decreases in demand for heating energy and increases in demand for cooling energy. This is projected to drive up overall electricity use and create higher peak demands in most regions, but it also may reduce the use of heating oil and natural gas in winter. Although the energy industry will be affected in multiple ways by changing weather patterns, it is generally considered to have the financial and the managerial resources to adapt (see also the discussion of impacts of climate change policies below).

Sea level rise and storm surges will increase the risk of major coastal impacts on vulnerable energy and industrial infrastructure and transportation (CCSP, 2007, 2008b), including temporary or permanent flooding of airports, roads, rail lines, and tunnels. More frequent extreme precipitation events would increase the risk of disruptions and delays in air, rail, and road transportation, as well as damage from mudslides in some areas. Increases in the intensity of strong hurricanes would lead to more evacuations, infrastructure damage and failure, and transportation interruptions (NRC, 2008c). Arctic warming will lengthen the marine transport season, while permafrost thaw on land will damage infrastructure and reduce the ice road season.

As experienced in Melbourne, Australia during a 2009 heat wave, an increase in extreme heat can limit some transportation operations (including airports) and cause pavement and track damage when heat compromises construction materials (NRC, 2008c). On the other hand, decreases in extreme cold can provide benefits such as reduced snow and ice removal costs and reduced snow-related road closures, as well as a potential decrease in snow- and ice-related traffic fatalities.

Locations, Systems, and Populations Will Be Affected by Climate Change Responses As Well As by Climate Change Itself

Impacts of climate change that may require adaptation by human and natural systems will not be limited to the direct effects of changes in temperature, precipitation, storm behavior, and sea level. Climate change is also likely to create indirect effects through the impacts of climate change policies. Examples include the effects that limits on GHG emissions may have on energy prices, technology choices, and both regional and institutional comparative advantages. In some cases, such indirect effects present a greater potential concern than climate change per se, and in many cases the literature on such indirect effects is more substantial than that on direct effects (CCSP, 2007).

Examples of impacts of climate change policies that might require an adaptive response include the following:

- If climate change policies emphasize reductions in GHG emissions, as expected, then regional economies dependent on fossil fuel production and use (especially coal) are likely to need to transition to different economic bases or rely on new technologies. Studies that examined the effects of policies that limit the impact of climate change on U.S. regions (e.g., Oladosu and Rose, 2007) have generally shown that aggregate economic impacts on coal-producing regions would be negative but could be quite modest, depending on how policies are implemented and how the regions respond.
- If climate change policies tend to favor land-intensive renewable energy alternatives, especially energy from biomass, then land areas devoted to natural resource preservation, forestry, agriculture, and ranching may have new opportunities for income generation. Some stresses could develop from competition between food and energy crops for land area, between resource preservation and biofuel production, and between energy crops and other uses of scarce resources such as water.
- Proposed climate change policies may raise energy prices as relatively inexpensive fossil energy sources are replaced by lower-emitting but more expensive alternatives. Analyses of the amount of the increase vary according to assumptions about such issues as ancillary benefits of a switch to alternative fuels and effects of policies designed to stimulate technological change. But some net costs to consumers are likely, which would affect relatively energy-intensive aspects of economies and societies, including costs of both transportation and electricity supplies.
- Climate change policies that alter the nation's portfolio of energy supply and use technologies will inevitably create economic winners and losers, although very little research has been conducted on this topic. Most likely to be affected are industries related to fossil fuels and the structure of the electric utility industry (Richels and Blanford, 2008). There are already signs that the reduction of GHG emissions is a factor in competition and economic health within the automobile industry (Levy and Rothenberg, 2002; Vance and Mehlin, 2009) and this could be a harbinger of impacts in other sectors as well.

Such effects are, in fact, only one aspect of complex interactions between adaptation and mitigation (Wilbanks and Sathaye, 2007). Finding ways to integrate mitigation and adaptation at the national level has been difficult; but at a local level, most decision makers and stakeholders contemplating climate change actions find it unwarranted to consider one strategy apart from the other (see also Chapter 5). A key issue is how mitigation and adaptation actions relate to each other. Some options offer complementarities and synergies, while some work at cross-purposes with each other. For example, increasing the efficiency and affordability of space cooling helps to extend the benefits of cooling to a wider range of the residents of warming settlements; at the same time, it also reduces requirements for electricity generation to enable those services. On the other hand, choices between growing biomass for energy production and growing biomass as a GHG sink, both of which can be mitigation strategies, relate in different ways to adaptation strategies. For instance, bioenergy production can add to challenges in adapting to water scarcity in some regions. In addition, growing biomass intended for long-term carbon storage can be complicated by climate change impacts on regional ecological systems, along with associated adaptive land use strategies.

Other possible impacts of climate change policies—not all of them negative—include effects on choices of energy production and use technologies, on environmental emissions, and on international energy technology and service markets (for additional details see *ACC: Limiting the Magnitude of Future Climate Change*; NRC, 2010c). Yet another issue is possible side effects of "geoengineering" options, should they be implemented. In general, geoengineering options intended to reduce the amount of solar radiation reaching the Earth—such as by creating a sulfate cloud in the atmosphere—would be virtually certain to affect vegetation growth and rainfall regimes, although the magnitude and geographic distribution of the potential effects are not well understood. Options intended to reduce current levels of CO_2 in the atmosphere would require extensive carbon storage in places such as underground geologic formations, a possibility that presents a different range of impact and adaptation concerns. In either case, both known and unintended impacts could require adaptations in response.

COMPARATIVE METRICS OF IMPACTS AND VULNERABILITIES

In order to determine tradeoffs between various climate change policies and actions, scientists have attempted to find objective measures for dangerous climate interference (as prescribed in the United Nations Framework Convention for Climate Change 1992) that might push the system beyond its adaptive capacity. To date, such an objective characterization of "dangerous" climate interference has not been developed. Nevertheless, a framework to consider global key vulnerabilities was developed for the Intergovernmental Panel on Climate Change (IPCC) Third Assessment Report (Smith

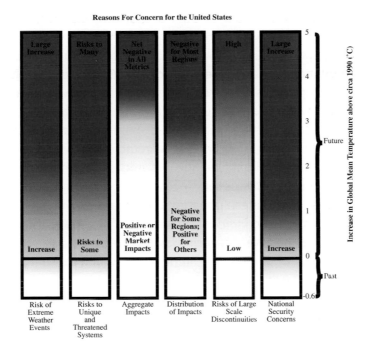

FIGURE 2.9 Risks from climate change for the United States. Climate change consequences for the United States are plotted against increases in global mean temperature (°C) after 1990. Each column represents country-specific outcomes associated with increasing global mean temperature for each of the six reasons for concern. The color scheme represents progressively increasing levels of risk: white indicates neutral or small negative or positive impacts or risks, yellow indicates negative impacts for some systems or low risks, and red means negative impacts or risks that are more widespread and/or greater in magnitude. Orange indicates a range of transition from risks calibrated in the modest risks of yellow and those calibrated in more severe and/or widespread risks of red. SOURCE: Yohe (2010); for details related to the assumptions in this figure see Appendix D.

et al., 2001; IPCC, 2001b; and updated in Smith et al., 2009a). Following Yohe (2010), the panel responded to Woolsey (2009), Peters (2009), and Burke et al. (2009) by adding a sixth "reason for concern" related to the national security interests of the United States. To be precise, the aggregate metrics, as applied to the United States in Figure 2.9, include the following:

1. *Risk of extreme weather events.* The likelihood of extreme events with substantial consequences for societies and natural systems such as increases in frequency or intensity of heat waves, floods, droughts, wildfires, or tropical cyclones, etc.

2. *Risk to unique and threatened systems.* The likelihood of imposing increased damage or irreparable loss to unique and threatened systems such as coral

reefs, tropical glaciers, endangered species, unique ecosystems, biodiversity hotspots, indigenous communities, etc.

3. *Aggregate impacts.* The likelihood of recognizing damages in aggregate. There are impacts distributed across the economy that can be aggregated into a single metric. Again, this reason for concern traditionally reported aggregate economic damages reports as, for example, the social cost of carbon.

4. *Distribution of impacts.* The likelihood of disparities of impacts (positive or negative) across regions or sectors. Some regions, sectors, or communities could face harm from climate change while others could even benefit. While this reason for concern historically focused primarily on economic metrics, recent work has aggregated subnational alternative metrics.

5. *Risks of large-scale discontinuities.* The likelihood of certain "threshold" phenomena that may have very large impacts. Examples include partial or complete deglaciation of the West Antarctic or Greenland ice sheets (that could lead to rapid increases in sea level), substantial reduction in the strength of the North Atlantic Meridional Overturning Circulation (that could result in a relatively rapid change in the climate system due to redistribution of heat in the oceans), and a "runaway greenhouse effect" (featuring more rapid warming) driven by methane emissions from melting permafrost.

6. *National security concerns.* The likelihood that growing attention to climate change risks and vulnerabilities that will occur beyond national borders will nonetheless require response by defense and other mission agencies within the U.S. government (for additional details, see Chapter 6).

It should be emphasized that this figure calibrates risks to increases in global mean temperature. It follows that the depicted transitions from low to high levels of concern do not necessarily reflect how risks might change at different rates of warming, nor do they necessarily indicate when impacts might be realized and how vulnerabilities might be influenced by alternative development pathways and the exercise of adaptive capacity. When applied to the globe and here to the United States, the underlying "reasons for concern" framework nonetheless continues to be a viable mechanism with which to describe key climate risks and thus help to identify priorities with regard to ongoing and future research initiatives and attractive foci for policy discussion and implementation (IPCC, 2001a, 2007a).

In summary, two qualitative conclusions emerge from the expert judgments that are embodied in Figure 2.9. Both reflect the evolving changes in climate variability that will be driven by long-term climate change.

1. If policy makers were provided with only aggregate economic metrics when they asked to be informed about the significance and timing of impacts, then they would miss many if not all of the other risks that are captured in five other equally appropriate reasons for concerns. Indeed, Figure 2.9 suggests that decision makers could, as a result, come to the erroneous conclusion that it may take quite some time for the country as a whole to experience the ramifications of "dangerous anthropogenic interference" with the climate system that has attracted the attention of many other countries. Of course, distinct localities and regions within the United States will have to cope with a diverse set of climate-driven vulnerabilities, and decision makers who work in these arenas will have an incentive to consider more focused economic aggregates. Even in these cases, though, interpreting aggregate economic indicators can be difficult, especially if those indicators ignore economic damages that will occur beyond specified borders—damages that are certainly part of a full and complete characterization of the potential economic risk to the nation.

2. Conversely, dangerous anthropogenic interference in the climate system will likely be discovered at all levels as climate change alters the intensities, frequencies, and regional distributions of extreme weather events. It is in these areas where investing in adaptive capacity and exercising adaptation options at the local level play their most critical roles; and it is through these manifestations that diversity in the climate risks facing various geographic regions scattered across the country and various climate-sensitive sectors scattered throughout the economy supports bottom-up approaches to evaluating adaptation needs.

The first conclusion is almost a corollary of the observation that aggregate economic estimates of damages too often ignore low-probability risks. Indeed, this deficiency is just one of a growing list of concerns about relying too heavily on monetary estimates—estimates that, for the most part, miss many nonmarket damages and nearly all consequences from social contingent consequences (see, e.g., Yohe, 2009a; Yohe and Tirpak, 2008). The second follows from the expectation that reasons for concern, by offering alternative but nonetheless aggregate metrics, communicate the diversity of those risks more effectively.

Care needs to be taken in interpreting these conclusions. Authors of the various versions of the reasons for concern have emphasized that they cannot be the sole basis of policy; too many assumptions and limits of knowledge are either buried or missing. It is important to understand, for example, that these six aggregate "reasons for concern" reflect adaptation only to the extent the capacity to respond is included in the underlying literature. It has long been understood that the capacity to adapt depends on development pathways that cannot be reflected in simple calibrations of changes in global mean temperature; it is now understood that the ability to exercise the capacity to adapt is very site specific. It follows that it was impossible, in this and other "embers" exercises, to include depictions of where existing or potential "coping ranges" might be exceeded by changes in global mean temperature.

Reasons for concern are best viewed as suggestions of where one might discover vulnerabilities and impacts that some or even most might consider "dangerous." Superimposed against ranges of temperature trajectories (as in Figure SPM-2 or SPM-3 in IPCC, 2001a), they might even suggest when such danger might begin to occur. It follows that reasons for concern, when properly applied, can help scientists and decision makers identify areas where more detailed analyses of vulnerabilities, and the associated opportunities for effective adaptation, might be most productive in directing research and informing policy design and implementation.

MAJOR SCIENTIFIC CHALLENGES IN ASSESSING CLIMATE CHANGE IMPACTS AND VULNERABILITIES AND THEIR IMPLICATIONS FOR ADAPTATION

At this early stage in analyzing adaptation needs and potentials, many scientific challenges remain in assessing vulnerabilities and impacts associated with climate change (*ACC: Advancing the Science of Climate Change*; NRC, 2010b). Six of the most significant of these challenges include:

1. *The level of scientific confidence in understanding and projecting climate change increases with spatial scale while the relevance and value of the projections for society declines.*

A branch of climate science called detection and attribution (D&A) seeks to understand the causes of observed changes in climate by comparing observed changes with those simulated by climate models under rising atmospheric GHG concentrations and against background climate variability in model simulations with no rising GHGs. For statistical reasons, D&A is most successful at large spatial scales. It was first used to identify human influence on globally averaged temperature over the 20th century.

Difficulties remain in attributing temperature changes on smaller than continental scales and over time scales of less than 50 years (IPCC, 2007b). Although the level of scientific confidence in climate change projections decreases at smaller spatial and temporal scales, the societal value increases. For example, while there is limited value in using the global averaged temperature for planning adaptation, information such as the projected ranges in possible changes in 100-year flood risk on a particular river at a given location can be very useful even if highly uncertain.

2. *A finer-scale understanding of climate change risks and vulnerabilities is needed.*

Impacts and adaptation are often local issues because the actual climate change impacts experienced will result from interactions of a specific climatic exposure with a specific population, sector, or system sensitive to that exposure, as well as the ability of that population, sector, or system to avoid, prepare for, and effectively respond to the risk. Thus, the same climatic exposure can have different consequences in different locations, and even in the same location at different time periods. Improvements are needed in the ability to project climate change at local and regional scales and to increase understanding of risks and the ability to design efficient and effective responses. Significant scientific challenges also remain in our limited understanding of the social, environmental, economic, institutional, and other factors that could interact with climatic changes to create impacts in any given location. Although Hurricane Katrina cannot be attributed to climate change, it demonstrated how hazard predictions with high certainties might fail to elicit proactive, necessary adaptations, partially due to a lack of understanding of local vulnerabilities and partially due to the difficulties of incorporating the latest natural and social sciences knowledge into practice and pre-disaster planning (NRC, 2006). Social groups particularly vulnerable to extreme weather events include the elderly, pregnant women, children, people with chronic medical conditions, people with mobility and cognitive constraints, and the urban and rural poor—all groups that are disproportionately represented within low-income communities (Balbus and Malina, 2009). Understanding such vulnerabilities in advance, and implementing appropriate strategies to increase resilience, affects the magnitude and extent of climate impacts (Kates et al., 2006; NRC, 2006).

Projected increases in the frequency and intensity of extreme weather events highlight the need to increase understanding of those most at risk, both today and in future societies with possibly different risk profiles. However, research on vulnerability and impacts (and associated sustainability indices) has frequently been at scales too large to incorporate social justice issues. One counterexample is heat waves, where research at finer scales has shown that poorer areas of cities often have higher temperatures because of fewer green spaces, and that the residents of these areas may have less access to air conditioning or may not open windows during heat waves for fear of crime, thus increasing vulnerability.

3. *Multiple stresses will interact with the impacts of climate change, leading to different vulnerabilities to the same climate condition in different locations and a need for different adaptive responses.*

Actual impacts will depend not only on climate but also on changes in other stresses over the same period. For example, impacts of climate change on vulnerable coastlines in 2070 will be shaped not only by changes in sea level, storm tracks, and storm intensities but also by land subsidence, changes in population size and distribution, economic activities and wealth, technology, and institutional structures. Understanding how interactions with these other factors accentuate or ameliorate climate change impacts is important for adaptation planning. For example, detailed projections of socioeconomic scenarios are often not available beyond several decades into the future (and if they are available, they are highly speculative), which limits our ability for integrated modeling to understand how interactions could play out over time.

4. *Adapting to changes in averages versus changes in extremes results in a fundamental scientific and policy challenge.*

Projections of the impacts of climate change tend to focus on changes in average weather variables, particularly changes in average temperature. As important as these changes are likely to be—for example, how increasing average temperature affects the suitability of particular cereal crops for a given region—the actions required for adapting to averages can be different from the actions required for adapting to extreme events. Strategies to manage the risks of climate change need to address projected increases in the frequency and intensity of extreme weather events, as well as unexpected threshold events. Depending on the cost of adaptation options versus the cost of impacts they are designed to avert, it may be helpful to prepare for low-probability/high-consequence events. Science and engineering needs include reevaluation of boundaries of flood plains, better flood maps, and redesign or retrofitting of hospitals and other critical infrastructure so that services would not be disrupted during an extreme weather event. Effective adaptation will thus require consideration of climate change risks along multiple dimensions: increasing resilience to warmer temperatures and average changes in the water cycle while at the same time increasing resilience to extreme weather events—and doing both while considering current and future changes in other driving forces.

5. *Interactions and integration across regions and sectors cause considerable complexity and will lead to unanticipated consequences of both impacts and adaptations.*

Climate change impacts in one sector or region usually spread secondary and tertiary impacts elsewhere. For instance, reducing impacts of summer warming on the quality of life of urban populations is likely to call for more air conditioning in homes and places of work. This will impact the energy sector by adding peak demands for electricity production, which can in turn impact the water sector by requiring more cooling water for thermal power plant operation. Likewise, agricultural adaptations in a region may call for increased use of irrigation, while water resources adaptation in that region may call for decreased water availability for irrigation.

Impacts can cross regional boundaries as well. For instance, more intense storms in vulnerable areas can mean flows of evacuees to other regions, along with at least temporary shortages (or increases in the price) of products and services disrupted by the storms, as was the case with energy products after Hurricane Katrina (Bamberger and Kumins, 2005). Further research is needed to improve our understanding of how to effectively develop cross-regional and cross-sectoral adaptation plans.

6. *The types of impacts, vulnerabilities, and adaptation options are different for natural and human systems.*

Both national and international assessments have described the broad patterns of recent and projected responses of natural systems and biodiversity to climate change (IPCC, 2007a; MEA, 2005; USGCRP, 2009). It is highly likely that most natural systems are sensitive to climate change. Much of our current understanding of ecosystem dynamics, however, is based on observations and models that assume less dramatic directional changes in environment and ecosystem dynamics. These models and theories provide an important starting point for understanding rapid change, but they will undoubtedly require reassessment as new patterns of environmental and ecological controls emerge.

These changes are likely to result in the loss of some ecosystems and the formation of novel ecosystems due to the loss of some species and arrival of others. Loss of biodiversity is quite likely, including both loss of rare species and loss or reduced importance of keystone species. During these times of rapid biological adjustment, metapopulation dynamics (i.e., interactions among partially isolated subpopulations) and migration of species across increasingly fragmented and human-modified landscapes are likely to exert greater influence on ecosystem structure and functioning than in the past. All of these changes could reduce the resilience of natural ecosystems and make them more vulnerable to threshold changes. These and other broad changes in the ground rules by which ecosystems operate create significant scientific challenges in understanding and predicting the patterns and consequences of changes in natural ecosystems.

The prospect of widespread ecological change also raises two pragmatic questions that represent additional research challenges: How can the rates of undesirable ecological change (e.g., loss of biodiversity) be minimized? And how can the flow of essential ecosystem services on which society depends be sustained in the face of rapid ecological change? Both questions will require improved understanding of the dynamics of social-ecological systems and collaborations.

ADAPTATION AND UNCERTAINTY

Adapting to climate change impacts will require doing our best to understand the factors that drive both the impacts and our ability to respond. This reality has led to urgent calls for more information about the range of possible impacts and the level of certainty in our projections of the future. It is clear that society cannot avoid the risks of climate change entirely. One challenge for decision makers will be the limits to our ability to identify and reduce uncertainties related to climate change.

Major uncertainties in determining future climate include the natural internal variability of the climate system, the trajectories of future emissions of GHGs and aerosols, and the response of the global climate system to any given set of future emissions (see also Chapter 4 and Meehl et al., 2007; NRC, 2010c). The magnitude and sources of these uncertainties can be explored using global climate models. These models have become more sophisticated and accurate over time in replicating the historical record. However, it is unlikely that climate models will be able to predict the future on fine spatial scales with a high degree of accuracy on long time scales. At best, climate models can provide insights about the range of possible futures.

Lack of certainty about future conditions is commonly, but often inappropriately, used as a rationale for inaction. In fact, improving our understanding of the kinds of uncertainties that we face will be helpful in risk-management decisions, even if the uncertainties cannot be readily quantified. For example, some uncertainties result from processes that are still missing from the climate models but are potentially resolvable in the future (e.g., changes in climate that result from changing land use and land cover). There are other uncertainties that are inherent in the complexity of the climate system itself, and it is unlikely that those kinds of uncertainty will be reduced significantly. For example, the uncertainty of the long-term trajectory of GHGs is very likely not to be reducible (CCSP, 2009c).

Another source of uncertainty comes from the fact that current global climate models operate at relatively coarse spatial scales (hundreds of kilometers or miles), and thus do not accurately represent conditions in specific places; rather, they represent average conditions across broad regions such as the entire Southwest.[2] This prob-

---

[2] However, it should be noted that the resolution of global models is increasing, and some of the simulations for the next IPCC report may be run at 50 km for the near-term future (next 20 years) .

lem of spatial scale is overcome through the application of various "downscaling" techniques—ways to generate information at higher spatial resolution from coarse-scale global model output. There are three primary methods of downscaling: (1) simple downscaling, where the coarse-resolution information is simply interpolated to higher resolution, or the coarse-scale changes in climate are used in the context of higher-resolution observed data; (2) statistical downscaling, which relies on statistical relationships between historically observed large-scale climate variables and local climate (e.g., daily temperature in a specific city) that are then applied to the climate change context; and (3) dynamical downscaling techniques, such as regional modeling, where a higher-resolution climate model is applied to just part of the Earth's surface (e.g., the western United States) and is "nested" in the global models.

Regional climate models can better represent smaller-scale processes such as those related to complex terrain (e.g., mountains) and provide data at scales closer to those at which decisions are made (within a watershed, for example). Regional climate models are useful in trying to understand the physical processes that control regional climate and the likely impacts of climate change within regions and sectors for risk-management planning. However, "downscaled" climate data can introduce other sources of uncertainty and are not yet the panacea that many resource managers hope they will be (Wang et al., 2004).[3] Making adaptation decisions in the context of uncertainty will remain a challenge, but one that can be overcome with careful attention to improving the understanding and characterization of—and the ability to communicate—the nature and sources of uncertainty.

CONCLUSIONS

The United States is already experiencing impacts of climate change that require adaptation. Some of these impacts are already testing, or soon will seriously test, the nation's coping mechanisms. In summary, the panel finds that climate change impacts are certain to increase throughout this century, requiring significant effort to adapt in order to avoid socially, economically, and environmentally disruptive changes in systems with high value to society. Adaptation options need to address current and projected changes in mean weather variables as well as increases in the frequency and intensity of many extreme events.

---

[3] Different regional climate models produce different responses to the boundary conditions from the global models, presenting another source of uncertainty. Also, high-resolution modeling must be considered in the context of the other uncertainties mentioned above. Nesting one regional model inside one global model, regardless of how high the resolution, will not provide important information about the larger-scale uncertainties, and can even be misleading and create a "false certainty."

Impacts later this century will be notably greater if GHG emissions are not stabilized at a moderate level. If the magnitude of climate change is relatively severe, as depicted in the USGCRP higher projection, then regions, sectors, and systems will be hard pressed to cope with impacts and their costs. In addition, impacts of climate change are highly diverse and disaggregated, in many cases playing out at localized geographic, sectoral, and societal scales. As a result, effective approaches to adaptation will likely vary from case to case.

In most cases, impacts are imbedded in interactions between climatic changes per se and other driving forces, such as changes in demographics, economics, land use, and technology, which also vary from case to case. Therefore, impacts and vulnerability are place-based and fundamentally driven by the scale at which the impact occurs. Many scientific challenges remain in assessing impacts and vulnerabilities and providing the specific and localized information needed to guide adaptation decisions.

> **Conclusion:** Many current and future climate change impacts require immediate actions to improve the ability of the nation to adapt. Because some impacts may not require immediate attention, possible adaptation options need to be prioritized based on where and when urgent action is needed. This highlights the need to identify vulnerabilities, impacts, and adaptation options across the nation, at all levels of decision making.

> **Conclusion:** Gaps in the knowledge required to link anticipated impacts with appropriate adaptation strategies and actions need to be addressed as a high national research priority.

> **Conclusion:** It is inadequate to provide policy makers with only aggregate economic metrics to convey the significance and timing of climate change impacts. Aggregated data miss most nonmarket damages and nearly all social contingent consequences that society might deem unacceptable, including those from outside our borders.

> **Conclusion:** Uncertainty about the nature of future climate change impacts in specific locations is not a rationale for inaction but a call for better understanding and communication of the sources and nature of uncertainty in the context of decision making.

CHAPTER THREE

What Are America's Options for Adaptation?

I f the United States is to cope effectively with the impacts of climate change, it will need an array of adaptation options to choose from. Unfortunately, adaptation to climate change has been a low national priority, and very little research has been devoted to identifying and evaluating options for adaptation. In the short term, the nation can draw lessons from past experience with adaptations to climate *variability*, limited experience with climate change adaptation already undertaken in some regions of the world, a limited number of careful analyses of adaptation possibilities, and from an onrush of creative thinking in connection with emerging efforts to do adaptation planning. But, in many cases, the options that we can identify for *adaptation to impacts of climate change* lack solid information about benefits, costs, potentials, and limits for three reasons:

1. *Attribution.* Climate change is just now emerging as a cause of impacts; therefore, it is difficult at this stage to document effects of adaptation in reducing those impacts.
2. *Diversity.* Which adaptation actions make sense depends very heavily on context: the nature of the impact, the geographical scale and location, and the sector(s) affected. As a result, general conclusions about effects of particular options are often difficult to support.
3. *Knowledge base.* Very little research has been carried out on climate change adaptation actions to date (as distinguished from determinants of adaptation capacity; see Chapter 5).

Society's need to cope with changing climate and environmental conditions is not new; people have been adjusting to their environment since the dawn of civilization. Agriculture is one of the earliest examples: over the ages, farmers have repeatedly adjusted cultivation practices and bred new plant and animal varieties suited to varying climate conditions. In recent times, the development of floodplain regulations, insurance, wildlife reserves, drinking water reservoirs, and building codes all reflect efforts to stabilize and protect our homes, livelihoods, and food supplies in the face of a variable climate. However, for the past 10,000 years, climate has been relatively stable, and weather patterns have fluctuated within a rather predictable range. Our growing awareness that the Earth's climate is changing, and that we are facing novel future climate conditions that will interact with and compound our current economic and environmental challenges, has created a new context and a sense of urgency for climate adaptation planning (Adger et al., 2009; Moser, 2009b; Rockstrom et al., 2009).

Adaptation measures now being considered include both extensions of past practices and novel strategies for addressing uncertainty and change. For example, newer efforts incorporate the necessity of anticipating a different climate and potential threshold events and conditions that will be outside the range of our past experience. The goals of our adaption efforts, however, remain the same as those in the past: to minimize harm and to take advantage of opportunities while sustaining human welfare and ecological integrity in the face of a changing environment.

Some attention to adaptation to climate change is already under way in sectors most likely to be affected, from agriculture to tourism, although information about such voluntary actions is limited and their effects will have to be evaluated over time. Most of the explicit adaptation planning is occurring now at state or local levels. Much of this planning has roots in the late 1990s regional assessments by the U.S. Global Change Research Program. Many of the state and local planning efforts have been supported by federal legislation, federal-state partnerships such as National Oceanic and Atmospheric Administration (NOAA)-sponsored Regional Integrated Sciences and Assessments (RISAs) and the Coastal Zone Management Program, and nongovernmental organizations (NGOs) such as the Center for Clean Air Policy (CCAP) and the International Council on Local Environmental Initiatives (now called the ICLEI-Local Governments for Sustainability; Chapter 5). Support from such diverse organizations indicates that the nation has considerable experience with planning at multiple scales and suggests that planning for climate change adaptation within the United States is likely to require coordinated public-private planning partnerships to span these scales. Significant adaptation planning for climate change has also occurred internationally (as illustrated in case studies in Chapters 5 and 6), stimulated by increasing awareness of climate change impacts and their serious societal and ecological consequences (IPCC, 2007a; Stern, 2007). In the United States, most adaptation planning at all scales has been initiated since 2005, and early efforts have largely focused on information gathering, vulnerability assessment, and organization—not yet on actions (Table 3.1). Therefore, despite increasing recognition of the urgent need to adapt to climate change, there is a very short history of past successes and failures from which to learn (Moser, 2009b).

This chapter provides examples of the range of options available for adapting to climate variability and extremes in key climate-sensitive sectors. The panel notes that adaptation to climate variability and change is an activity whose depth and breadth

TABLE 3.1 Early adaptation activities

Urban Leaders Adaptation Initiative Partner	Example Adaptation Activities
Chicago, Illinois	Developed Chicago Climate Action Plan in 2008; developed vulnerability and economic impact analyses; prioritized planning strategies to address impacts; raised substantial external funds to support adaptation programs; conducted downscaling of climate information for local decision making.
King County, Washington	Established the "Ask the Climate Question" approach to adaptation; funded a district-wide study of implications of climate change for water quality and quantity; worked with the Climate Impacts Group (CIG) at the University of Washington to conduct an infrastructure assessment and develop a Geographic Information System tool; in partnership with CIG developed the handbook *Preparing for Climate Change: A Guidebook for Local, Regional and State Governments*; implemented changes in water reclamation and distribution to expand municipal wastewater reuse.
Los Angeles, California	Established a Climate Adaptation Division within the Environmental Affairs Department and a Director for Climate Adaptation; developed downscaled regional climate information for decision making; explored urban heat island effects and prioritized areas to receive shade trees; adopted the Los Angeles Green Building Ordinance.
Miami-Dade County, Florida	Used Federal Emergency Management Agency (FEMA) funds to strengthen buildings and develop hurricane shelters; engaged 250 stakeholders from multiple backgrounds and sectors and established the Climate Change Advisory Task Force; developed a report on adaptation strategies for the built environment and recommended developing minimum criteria standards for public investment. Working as a member of the Florida Climate Change Adaptation Technical Working Group, released a report to the governor on policy recommendations.
Milwaukee, Wisconsin	Preparing for more intense flood events, Milwaukee planners are aiming for a target of zero stormwater overflows per year to protect water quality in Lake Michigan; have constructed a deep tunnel for increased stormwater storage and conducted an analysis on stormwater infrastructure investments; and examined existing development codes to determine ways to encourage green spaces including rain gardens for increased infiltration. They are also working with other state partners on downscaling climate information and identifying adaptation strategies.

continued

TABLE 3.1 Continued

Urban Leaders Adaptation Initiative Partner	Example Adaptation Activities
Nassau County, New York	Nassau County recently completed its first Multi-jurisdictional Hazard Mitigation Plan, funded by the FEMA Pre-disaster Hazard Mitigation Program. This has identified a series of measures to reduce disaster impacts and encourage smart growth to avoid impacts of flooding, storm surge, and sea level rise.
Phoenix, Arizona	Phoenix has incorporated climate change adaptation actions into the city's sustainability program. This program focuses on land use, pollution prevention, and water-use measures that increase climate change resilience. The Phoenix Water Resources Plan includes long-term projections of water supply and demand that incorporate assumptions about changes in regional water supply availability. They have created an interdepartmental task force to address urban heat island issues in the urban core, including assessments of changes in building materials.
San Francisco, California	San Francisco has created a comprehensive climate action plan aimed at mitigating greenhouse gas emissions and understanding climate impacts, with a particular focus on environmental justice issues. San Francisco worked with multiple other large water utilities to create the Water Utility Climate Alliance, which now represents more than 40 million people in the United States and has been working to identify research needs in support of decision making.

SOURCE: Information from CCAP (2009).

vastly exceed its profile in the academic literature because the intended outcome is a practical, not an academic, result. Where possible, the available literature is cited, but the examples given below of possible adaptation options include some that have been widely and successfully tried but not discussed in the literature, as well as some that are novel or have been frequently suggested but never tried. Space precludes a thorough discussion of the history and practice of the various options presented below.

The chapter follows with an examination of lessons that can be learned from a suite of integrated climate change adaptation planning processes under way in the United States and elsewhere. From these sectoral elements and lessons learned from early case studies, the panel summarizes findings that can provide a basis for designing climate change adaptation strategies and plans. These lead toward several recommended steps that can be implemented immediately or very soon (Chapter 8).

SECTORAL ADAPTATIONS TO CLIMATE CHANGE

Most current adaptation plans represent targeted efforts to address vulnerabilities in a single sector. They often build logically on past programs that have dealt with variability and extremes, such as extreme drought in agricultural areas, heat waves, or disease outbreaks in cities. This section summarizes possible options for adapting to climate change that have been identified in each of a number of sectors, including long but not exhaustive lists of ideas as illustrations of current perspectives and knowledge. As noted above, many of these options have not only not yet been tested and proven effective as adaptation options to climate change, but in most cases their benefits, costs, potentials, and possible limitations have not been carefully analyzed. However, they do represent a range of ideas about potential options to reduce vulnerabilities that are currently being discussed.

The "sectors" that the panel selected for analysis are sensitive to climate change and provide examples of the types of issues that are frequently managed by a single agency (e.g., agriculture, transportation, energy), are focal climate-sensitive public concerns (e.g., ecosystems, water, health), or are regions that face a consistent suite of interrelated issues (e.g., coastal zones). In general, these are sectors with great reasons for concern and are considered a high priority for adaptation. The panel also identified the policy level or agency—federal, state, local/city, private sector, NGO, or individual citizens—best poised to implement each option. In many cases, adaptation options will be implemented across scales and with multiple "actors." The adaptation options in the tables that follow are either examples from the literature or are based on expert judgment by members of the panel. Some are demonstrated responses to past incidents of climate variability such as flooding or prolonged drought. The suitability of any option generally depends on temporal and spatial context, as described in the cited references. Consequently, the adaptation options listed in the tables should not be construed as universally applicable recommendations. Instead, this panel stresses the importance of weighing the costs and benefits of each adaptation option on a case-by-case basis in the context of local needs, conditions, and impacts on other sectors.

Ecosystems

Increasing atmospheric concentrations of greenhouse gases (GHGs) and associated climate changes directly affect ecosystems and the benefits they provide to society (i.e., *ecosystem services* such as food, fiber, regulation of water quantity and quality, and the cultural, aesthetic, and recreational benefits of ecosystems; also see Box 2.1) (MEA, 2005). Climate change also affects ecosystems through its impacts on underlying ecosystem conditions such as soil fertility, species composition, and disturbance patterns (NRC, 2008a; USGCRP, 2009). Some of the greatest changes in ecosystems, however, are being driven largely by changes in land use, nutrient and other contaminant additions, loss of key native species, invasions of exotic species, and other human-caused disturbances (Foley et al., 2005). Some of the most important impacts of climate change on ecosystems will result from interactions with these other human-caused impacts on ecosystems (USGCRP, 2009).

Managers, particularly at state and local levels, generally have considerable experience with adaptation actions that yield immediate benefits under climatic extremes—for example, managing for water-conserving species and storing water to maintain adequate environmental flows during droughts. Similarly, the risk of ecosystem degradation in response to climate change can be reduced by managing other human-generated stresses such as pollution of freshwaters, estuaries, and coral reefs or rangeland erosion induced by overgrazing (Table 3.2). In some cases, such as ocean acidification, there is no known adaptation option other than to reduce rates of change in GHG concentrations and climate.

Along with the development of a means of pricing and accounting for ecosystem services, sustaining ecosystem benefits for society over the longer term will require novel approaches such as periodic groundwater recharge during times of water surplus, filling of canals to prevent saltwater intrusion, and collaboration with stakeholders to co-manage the fringe of suburbs and other human developments surrounding many conservation lands (see additional examples of ideas in Table 3.2). Development of such long-term adaptation strategies will require experimentation (adaptive management) at appropriate scales and engagement of government at all levels, as well as the private sector, NGOs, and individual citizens (West et al., 2009). Government actions are important in aligning incentives with adaptation goals, particularly over the long term, and in facilitating a nationally coordinated effort by specifying minimum standards and/or providing funding opportunities (Adger et al., 2009). Maintaining a diversity of options by sustaining biodiversity and encouraging diverse management approaches at all scales will provide the nation with the necessary flexibility and resilience to respond to uncertain future climate changes (Chapin et al., 2009; Folke, 2006; West et al., 2009).

Indirect effects of climate change, operating through changes in species composition and natural disturbance regimes such wildfire, insect outbreaks, and disease, are less certain but will likely have greater impact on ecosystems than direct effects such as temperature changes (MEA, 2005). The United States has considerable management experience at local to national levels in protecting endangered species, controlling the spread of exotic species, and managing natural disturbances, and this experience provides a fundamental starting point for dealing with indirect effects of climate change. Despite current management efforts, however, recent climate warming has already begun to affect disturbance patterns and the well-being of native and exotic species in many parts of the nation, indicating that our current knowledge and practices are insufficient to address these issues over the long term (Brooks et al., 2004; Westerling et al., 2006). New strategies might be needed to reduce risk (for example, through redundant refuges and no-take zones to reduce pressure on protected species or overharvested stocks); to manage for migration of desirable species and barriers to weedy species; and to manage rare or novel ecosystems for resilience and ecosystem services, including cultural and aesthetic value, rather than for past species composition (Hobbs et al., 2009; West et al., 2009) (Table 3.2). Experimenting, modeling scenarios of alternative futures, and engaging stakeholders in transparent decision-making processes will be critical to developing long-term adaptation strategies that can successfully address the indirect effects of climate change (Carpenter et al., 2009; MEA, 2005).

Managing ecosystems for adaptation to climate change also requires more consistent use of currently recognized best practices such as monitoring change, managing for multiple ecosystem benefits, and keeping disturbance at acceptable scales; attention to climate interactions with other processes such as wildfire and species invasions; and managing ecosystems as coupled social-ecological systems in which society both responds to and affects ecosystems and the climate system.

Agriculture and Forestry

Agriculture and, to a lesser extent, forestry have developed well-proven methods to adapt to direct effects of climate stress in the short term. These include changes in cropping, planting, and harvesting practices, as well as breeding and seed collection programs that provide genetic types and varieties of plants and animals adapted to different water and temperature conditions (USGCRP, 2009) (Table 3.3). Economic constraints and uncertainties about weather and global markets, however, can limit successful implementation of these approaches (Easterling et al., 2007). The application of these traditional approaches is particularly challenging with long-lived crops, including fruit and forest trees, which may experience considerable climate variability and change over the years between germination and maturity (Millar et al., 2007). Long-term adaptations to climate change may require development of new varieties, genetically engineered crops, use of different seed stocks to sustain a diversity of genetic stocks, or a shift of agriculture to different regions as climate in those areas becomes more favorable for certain crops or livestock. Other adaptation options might include development of irrigation techniques that use less water, switching from rainfed to irrigated agriculture where groundwater pools can be sustained, or, in the worst cases—where neither precipitation nor groundwater is adequate—possible cessation of agriculture (California Department of Water Resources, 2008; CCSP, 2008c; Easterling et al., 2007; NRC, 1996a).

Short-term responses to climate-induced increases in pests and diseases can involve improved pest management and, in the case of agriculture, application of integrated pest management practices, including development of pest-resistant varieties, use of herbicides and pesticides, and maintaining habitat for natural predators of pests. Over the longer term, technologies such as remote sensing of pest outbreaks may provide a valuable early warning system, and landscape management changes may reduce the potential for spread of pests and diseases or the spread of forest fire (see Table 3.3).

There are potential tradeoffs and synergies between agricultural adaptation and adaptations of the water and ecosystem sectors. Increased irrigation in response to drought, for example, competes with natural ecosystem flows and domestic water needs. Pest management must be carefully targeted to prevent elimination of natural predators of pests that reside in natural ecosystems. Synergies can also be developed, such as taking advantage of the diversity of predators in natural ecosystems as a component of integrated pest management in agriculture and forestry.

Water

Because of the widespread occurrence of both chronic and periodic water shortages, many potential adaptation strategies have been developed and tested for variations in water availability, from building dams to encouraging conservation by households and other water users (Table 3.4). Although most attention has focused on adaptation options that provide immediate benefit, the severe societal consequences of water scarcity have also spurred several innovative adaptations designed to provide greater long-term benefits, despite substantial initial costs (Table 3.4). Engineering approaches such as dams and water delivery infrastructure, underground (aquifer) storage and recovery systems, and seawater desalination plants have been developed and implemented throughout the country in response to historic climate variability. Conservation adaptations, including changes in behavior as well as water-saving technologies, are available in all water-use sectors and are generally the most cost-effective options, especially where synergies between energy and water conservation can be achieved.

Though increased flooding as a consequence of climate change has generally received less attention to date than drought impacts, multiple "standard" engineering approaches are available to reduce short-term flood risk (levees, "hardening" of coastlines), as well as a host of more innovative and environmentally friendly options (see the section "Coastal Area Vulnerabilities" in this chapter; Beatley, 2009). In cities, flooding associated with storms and sea level rise will challenge the capacity of storm drains and water and wastewater treatment facilities to handle increased flows and in extreme conditions may threaten the integrity of water supplies. Short-term adaptations can include actions that improve current functions, such as repairing leaks in wastewater and water supply lines and infrastructure improvements to match flow capacities to projected changes. Longer-term adaptation strategies could include development of distributed supply and treatment nodes that are less vulnerable to disruption and require smaller water volumes.

The effects of climate change on water quality and temperature have received less attention than effects on water quantity; as a result, both short- and long-term adaptation strategies will require substantial development and testing to determine their effectiveness. Key temperature-related impacts include threshold conditions in lakes, reservoirs, and rivers for temperature-sensitive fish and other aquatic species, as well as implications for water used for cooling in industrial plants and energy production. Adaptation approaches generally involve changes in reservoir operations to manage downstream temperatures. Water quality adaptations often involve changes in land use, such as development of vegetated buffers to protect waterways from sedimentation during floods, or non-point source and point source pollution management systems that can withstand high flows.

Little attention has been given to the possibility of modifying water rights laws and practices in the United States in response to changes in water availability, although this has proven necessary and effective in addressing chronic and increasing water shortages in South Africa and Australia (Chapter 5; Carpenter and Biggs, 2009). Water rights allocation is within the purview of the states, and each state has a different approach to managing water supply availability. The institutional complexity of water rights administration and strong incentives for maintaining the legal status quo have limited the interest in more flexible management systems. Nevertheless, greater flexibility in water management may be an appropriate short- and long-term option in many cases.

Water allocations to agriculture, ecosystems, energy, recreation, transportation, and domestic use present both tradeoffs and synergies. Tradeoffs are inevitable when water is insufficient to meet the needs of all users, which is the rule rather than the exception even in the more humid regions of the United States. For example, dams that are built to provide water storage for agriculture and domestic use may negatively impact endangered species or the services provided by undammed streams and riparian zones. The recent (and unexpected) drought conditions in the southeastern United States provide an excellent example of the relationships between sectors: a high-cost impact of the drought of the early 2000s in Alabama and Georgia included limits on energy production caused by water supply shortfalls. More positive synergies can occur when intact ecosystems are used to buffer flows and filter contaminants that might otherwise threaten public water supplies.

Health

Health concerns due to climate change are already evident and require adaptation. Concerns include increased frequency of climate-related extreme events (e.g., heat waves, waterborne diseases, and wildfire smoke) and climate-induced ecological changes (food- and waterborne diseases, and changes in distribution of diseases and their vectors) (Table 2.2, Box 2.2). Proposed adaptation options for the short term focus on altering and augmenting current public health programs and activities to increase their effectiveness in a changing climate (Table 3.5). These include early warning systems, emergency response plans, and public outreach for extreme events. Such programs were designed assuming the current climate would remain essentially constant, so they often need adjustment to address increasing risks from extreme weather events. Many early warning systems, for example, were not designed for monitoring and evaluating in a changing climate.

Reducing health risks related to climate change over the longer term may require new decision-support tools and changes in other sectors that affect public health, such as urban design to minimize the urban heat island effect through greater use of trees and green spaces. Education and training programs for health care professionals have been identified as important adaptation options to build capacity to address climate-related health needs, including postdisaster mental health needs (Frumkin et al., 2008; Jackson and Shields, 2008).

Climate change and associated extreme events such as floods and hurricanes will change the locations and frequency of disease outbreaks caused by water-, food-, and vector-borne pathogens (Ebi et al., 2008; Frumkin et al., 2008; Jackson and Shields, 2008). In the short term, adaptation can be facilitated by new early warning systems, vaccines, and upgrading of water treatment (both supply and sewage) facilities in disaster-prone regions, as well as public education campaigns to increase awareness of new disease risks. Health risks can be reduced by managing water supplies to reduce flood risk (see section "Water" in this chapter) and by managing ecosystems to eliminate breeding sites of insect and other vectors and to reduce the spread of allergenic plants (see section "Ecosystems"). Pathogen and disease vector surveillance and management programs can provide early detection and enhance our ability to take action to limit risks.

Both short- and long-term adaptation will require institutional changes in public health programs, training of health care professionals, and public awareness. For example, federal leadership for health organizations and agencies could facilitate collaboration and coordination on development and implementation of new early warning systems and decision-support tools. In addition, health-related climate considerations must become a more integral component of urban planning and ecosystem management (see sections "Ecosystems" and "Transportation").

In the public health sphere, synergies with other societal goals are generally stronger than tradeoffs, and most actions taken to cope with climate change impacts will improve health generally (no-regrets options) by, among others, improving the capacity to meet drinking water and other standards and reducing the likelihood of vector-breeding areas developing near communities during times of rapid ecological change (e.g., tropical deforestation and other land-clearing activities). Tradeoffs are primarily economic and can be minimized by making cost-effective choices in adapting health programs to deal with the additional stresses of climate change.

Transportation

Substantial engineering options are already available for strengthening and protecting transportation facilities such as bridges, ports, roads, and railroads from coastal storms and flooding to achieve short-term and long-term adaptation (Table 3.6) (NRC, 2008c). Infrastructure can be elevated, built stronger, protected by levees or dikes, and/or moved. For example, several of the major Gulf Coast highway bridges destroyed by storm surges during Hurricane Katrina have been redesigned and replaced by new bridges elevated well above anticipated future storm surges. Because most transportation systems are designed to last for decades, it is important for transportation planners to incorporate climate change in the planning and design cycle for such infrastructure (NRC, 2008c). Indeed, the Federal Highway Administration already encourages and funds the inclusion of climate change in metropolitan planning activities.

The general research approach for adapting to climate change is well established, although effective adaptation will require continued research and application. For example, there is a history of research on developing paving and other materials that are more heat resistant and on construction practices that protect permafrost. Research on climate or weather impacts on various modes of transportation has focused on extreme events (e.g., fog impacts on aviation and ice impacts on highways) and much of this research may be applicable to longer-term climate change issues.

Water-based transportation (e.g., Great Lakes Canal System and Mississippi, Ohio, Colombia, and Yukon rivers) may be constrained in some regions where runoff from the watershed declines and/or water demands for agriculture and cities increase. Adaptation may require redesign of ships and barges, a shortened shipping season, or a shift to alternate modes of transportation. Conversely, more extreme inland storms may well result in greater flood frequencies and levels. Revision of Federal Emergency Management Agency (FEMA) flood maps to reflect the probabilities of greater storm and flood events is critical to the construction of resilient structures along these inland waterways. Retreat of Arctic sea ice will likely open a new transportation corridor between the Atlantic and Pacific oceans. This may require new ship designs to deal with seasonal sea ice, development of new harbors, and development of new technology to handle fuel spills in water with sea ice.

Planning for climate change adaptation in the nation's transportation infrastructure will require new approaches to engineering analysis, especially the use of probabilistic analysis (i.e., risk analysis based on uncertain climate changes). New engineering standards will need to be developed to reflect future climate conditions and to be phased into ongoing rehabilitation projects where practical.

The nation currently has no experience in planning for or deciding when or how to abandon exposed coastal areas or communities that can no longer be adequately protected from rising sea level and greater storm surge. In vulnerable coastal areas, transportation systems and the people they serve will be placed at risk, and the social and political dimensions of relocation will provide a major research and policy challenge in adapting to future climate change.

Energy

Adaptation to impacts of climate change in the energy sector includes impacts on both energy production and energy use, many of which are focused at a regional scale. In terms of use, most of the research to date has been focused on energy use in buildings. Research results indicate that the demand for cooling increases 5 to 20 percent per 1.8°F (1°C) of warming, and the demand for heating drops by 3 to 15 percent per 1.8°F (1°C) of warming, depending on assumptions about the rate of market penetration of new technologies (USGCRP, 2009). Because nearly all of the cooling of buildings is provided by electricity use, while the majority of heating is provided by natural gas and fuel oil, the projected changes imply *increased demands for electricity*, especially where the population is growing and where relatively little space cooling has been needed in the past (CCSP, 2007; USGCRP, 2009).

Even as electricity demand increases in many regions, climate change may *reduce electricity production in some regions*. For example, climate change is likely to reduce water available for hydroelectric power in parts of the western United States that depend on meltwater from winter snows, and power plants located in areas vulnerable to more severe storms are at risk, especially from flood hazards. In addition, seasonal water shortages in many regions could threaten supplies of cooling water for thermal power plants, and higher air and water temperatures reduce the efficiency of power plant operations by a margin that is small for an individual power plant but can add up at a regional scale. For example, a major heat wave in Europe in 2003 required the temporary shutdown of several nuclear plants to prevent overheating. Moreover, the potential to generate electricity from other renewable energy sources besides hydropower, such as wind power and biomass, is also likely to be affected by climate change, although these effects are not all well understood (CCSP, 2007; USGCRP, 2009).

A somewhat different issue is the vulnerability of coastal energy facilities, especially in the Gulf Coast region, to combined impacts of sea level rise, more intense storms, and land subsidence (CCSP, 2008a). In the near term, adaptations are likely to emphasize protecting coastal energy and industrial infrastructures with barriers, but for the longer run, investment strategies for new infrastructures may consider shifts in location to less vulnerable areas.

Existing and emerging technologies are available to support a multipronged approach to adapting to anticipated climate-induced changes in the energy sector. The key first step is to reduce overall energy demands through increased energy-use efficiency and other steps; see *ACC: Limiting the Magnitude of Future Climate Change* (NRC, 2010c) (Table 3.7). This illustrates the importance of combining efforts to limit the magnitude of climate change with adaptation strategies in a single, integrated, climate choices program. In addition, engineering solutions are needed to make electricity generation less sensitive to climate change, through use of less-climate-sensitive tech- nologies and reduced water requirements (Table 3.7). Finally, emerging technologies could reduce climate sensitivity through the use of an integrated smart transmission grid to adapt to changing energy availability and demand. Renewable energy sources such as solar, wind, and water are periodic, episodic, or ephemeral. As an adaptation strategy, renewable sources that are widely dispersed could be connected through a smart grid and managed as a base load that would be supplemented by fossil energy generators. The Western Interstate Energy Board and the Western Electricity Coordi- nating Council have considered how such a grid could be orchestrated.

Although continued research will certainly improve U.S. capacity to adapt in the energy sector, current understandings of climate trends and current technologies can enable considerable progress in adapting the energy sector to climate change. In many cases, reductions in costs and increased reliability of electricity that would be achieved through these adaptive actions are likely to provide co-benefits, regardless of the magnitude of future climate changes.

Coastal Area Vulnerabilities

The key issues for adaptation to climate change in the coastal zone are the erosion and flooding that will become more pronounced with sea level rise, along with expo- sure to more intense severe weather events in some regions, all combined with land uses that are often extensive and valuable. In cases where these effects are projected to become more severe as climate change intensifies, a spectrum of potential adapta- tion options that vary in effectiveness over time are available (Table 3.8). For example, incentives that reduce development pressures and foster conservation in low-lying hazardous areas and encourage development in areas well above projected sea level rise would be advantageous in both the short term (reducing vulnerability to ex- treme storms) and the long term (reducing inevitable long-term losses). In contrast, strategies such as armoring of coastlines may enhance erosion in other locations, and levees may encourage development in flood-prone areas—results that are likely to be maladaptive over the longer term. Conservation in low-lying areas also provides opportunities for natural migration of coastal ecosystems in response to sea level rise. Over time, incentives for relocation to less vulnerable sites could allow gradual move- ment of communities to less hazard-prone areas and reduce the likelihood of costly disasters.

Coastal area vulnerabilities are a "sector" in which adaptation efforts have been especially active in recent years. In 2006, nearly two-thirds of the coastal states reported to NOAA that "coastal hazards" were a high priority and developed new 5-year strategies to address flooding, shoreline erosion, and coastal storms through their Coastal Zone Management Act (CZMA) programs (NOAA, 2006; Chapter 5). Since all of these hazards are expected to intensify in future climate scenarios, even where states have not yet engaged in formal "adaptation" initiatives, coastal management programs continue to improve long-standing policies related to the projected impacts of climate change. According to a 2008 survey, 4 state coastal programs had already developed a coastal adaptation plan or strategy, 7 states had a plan under development, and at least 12 other states anticipated launching new coastal adaptation planning efforts over the next few years (CSO, 2007, 2008). States were using a range of scenarios and methods, but the review of these initial planning efforts revealed six initial categories of adaptation policies focused primarily on the impacts of sea level rise: public infrastructure siting, site-level project planning and design, wetland conservation and restoration policies, shoreline stabilization, setbacks, and relocation policies; encouraging adaptive development designs (e.g., additional flood-height tolerance); and incorporation of climate change adaptation into other state, regional, and local plans.

The states also identified a number of key information needs for coastal adaptation planning, including high-resolution topography and bathymetry (CSO, 2007, 2008). Coastal programs were increasingly utilizing federal CZMA Enhancement Grants to study and plan for climate change impacts; however, funds available under this program may not be sufficient or intended to implement large-scale or long-term adaptation programs and are needed to address other program areas, such as the siting and review of energy infrastructure on the continental shelf. States have also engaged in coastal adaptation planning through other state and federal programs, including those of other NOAA branches, the Environmental Protection Agency (EPA), the U.S. Army Corps of Engineers (USACE), FEMA, and the U.S. Geological Survey (USGS). For example, the EPA is now funding pilot studies in eight estuaries under its "Climate Ready Estuaries Program."[1]

Other federal, state, and local programs for coastal areas are also beginning to incorporate sea level rise considerations in ongoing planning and program implementation. For example, USACE issued directive 1165-2-211 in July of 2009, requiring that all USACE civil works projects include an analysis of the impact of sea level rise using both the National Research Council report *Responding to Changes in Sea Level: Engineering Implications*, and Intergovernmental Panel on Climate Change (IPCC) projec-

---

[1] *http://www.epa.gov/cre/.*

tions (NRC, 1987; USACE, 2009). The guidance suggests using historic sea level rise as the low projection and then developing both intermediate and high estimates. The three projections thereby provide worst- and best-case scenarios on which to base engineering decisions. The Alaska and Gulf Coast Shoreline case studies (Boxes 3.1 and 3.3, respectively) illustrate the need for engagement in coastal issues by numerous agencies and programs (such as Emergency Management and Transportation departments), NGOs, and the private sector, as well as the difficulties in developing an integrated approach to coastal adaptation.

LESSONS FROM INTEGRATED CLIMATE CHANGE ADAPTATION PROGRAMS

Although human-caused climate change has been recognized as a distinct possibility for more than a century, and has been increasingly documented as a real and accelerating phenomenon in the last quarter century (IPCC, 2007a), surprisingly few climate change adaptation plans have been implemented anywhere in the world. Most of the climate change adaptation plans reviewed in this report, including some of the most advanced in the world, are still in the planning phase, with concerted planning initiated primarily in the past 5 years (since 2005). Nevertheless, one can draw lessons from existing adaptation activities, such as in Alaska (Box 3.1).

Many of the recent planning efforts at national to local levels encompass multiple sectors and suggest lessons that could provide useful starting points for designing a national strategy for climate change adaptation. Based on case studies presented in this section and elsewhere in the report (Figure 3.1), the panel has synthesized a number of key features that have proven valuable in climate change adaptation programs—or whose absence has seriously compromised the success of those programs.

Urgency of Climate Change Planning and Institutional Readiness for Implementation

Development of climate change adaptation plans has occurred most conspicuously in regions around the globe that currently experience severe climate change impacts (e.g., Alaska, Australia) or where leaders are concerned about vulnerability to impending impacts, particularly of sea level rise (e.g., Pacific Island States, Bangladesh, Maryland, Florida, California, and New York City). Alaska, for example, is warming at twice the global average rate, and several of the coastal communities eroding into the Bering Sea are already actively planning relocation to areas less threatened by coastal erosion (Box 3.1) (Hinzman et al., 2005; IAW, 2009). Similarly, Australia, which has experienced a decade-long drought, has taken dramatic actions to reduce water withdrawals and restructure water rights and has reorganized its agricultural enterprise to be better positioned to take advantage of climate opportunities and reduce climate risks (Garrick et al., 2009; Chapter 5). Maryland, New York City (Chapter 4), and Bangladesh (Chapter 6) are developing plans to accommodate sea level rise, and California (Chapter 5) and the entire lower Colorado River region (the states of Arizona and Nevada as well as Mexico; Chapter 6) are planning adaptations to reduce the impacts of increasing water scarcity (Feldman et al., 2008; NRC, 2007b).

These climate change adaptation plans are consistent with both recent climate trends and future projections (USGCRP, 2009). The co-location of adaptation planning with areas of observed or perceived impending risk suggests that scientific studies that identify such vulnerability hot spots (USGCRP, 2009) and outreach efforts to leaders and the public in these regions (e.g., RISA efforts in King County) could stimulate local-to-regional adaptation planning where it is most urgent. On the other hand, other vulnerable regions such as the U.S. Gulf Coast (Box 3.2) have been slow to develop comprehensive climate change adaptation plans, indicating that vulnerability, by itself, does not necessarily lead to comprehensive adaptation planning.

The success of climate change planning efforts varies considerably, depending on the complexity of tasks undertaken and institutional capacity to support the actions needed. In Philadelphia, for example, attention targeted to a single climate change issue (heat waves) that built on an existing institutional structure has allowed relatively rapid design and implementation of an appropriate adaptation plan (Box. 3.3) (Ebi et al., 2004). In contrast, development and implementation of plans to move away from the coast to escape storms, erosion, and sea level rise are extremely complex and often controversial (see Boxes 3.1 and 3.2). In Alaska, a consortium of state and federal agencies has been unsuccessful in implementing community relocation despite community and agency consensus on the necessity of moving and availability of funds to initiate the process. Inappropriate institutional structure and lack of adaptive capacity have been the major stumbling blocks (Box 3.1) (Bronen, 2008). In contrast, Australia has made substantial progress in revamping its complex agricultural program through innovative institutional changes that address climate change in an integrated fashion (Howden, 2009; Howden et al., 2007).

In summary, high vulnerability to observed or perceived future climate changes often, but not always, stimulates planning for climate change adaptation. The reasons why adaptation planning emerges in some vulnerable regions but not others are unclear, but they may relate in part to leadership and public awareness. Research is urgently needed to identify factors governing readiness for adaptation planning. Successful

> ## BOX 3.1
> ### Alaska
>
> Alaskan coastal and river communities have always experienced erosion and flooding, but climate change and infrastructure development have exacerbated these risks. Climate-induced changes already observed include (1) increased storm activity; (2) reduced sea ice extent, which increases the intensity of storm surges; (3) increased windiness; and (4) thawing of permafrost, which increases coastal susceptibility to erosion (ACIA, 2005; Hinzman et al., 2005; Huntington, 2000). Because of these increased risks, six Alaskan communities are in various stages of planning some type of relocation.
>
> Traditionally, many of these communities were seminomadic, moving inland during periods of severe storms and erosion (Marino, 2009). During the past hundred years, however, this resilience has been reduced by the building of permanent infrastructure such as schools, associated with compulsory education; airports and barge facilities to provide transportation access; and permanent houses with fuel and water supply. The U.S. Army Corps of Engineers has identified 160 additional villages in rural Alaska that are threatened by climate-related erosion, with relocation costs estimated at $30 to $50 million per village (IAW, 2009).
>
> In September 2007, the Governor of Alaska established an Alaska Climate Change Sub-Cabinet to prepare and implement an Alaska Climate Change Strategy[1] to address issues of adaptation and risk. The Immediate Action Workgroup (IAW) sketched out an emergency suite of projects that could be completed within 12 to 18 months to protect life and safety in the six communities requiring immediate relocation. The IAW also identified the governance issues that needed to be resolved in order to fully implement a relocation strategy for communities at risk. The Alaska legislature appropriated $8.3 million and leveraged $31 million in federal funds in fiscal year 2009 to protect current infrastructure with revetments, but no funds had yet been appropriated to begin the relocation process (IAW, 2009).
>
> Barriers to relocation include disagreement within some villages about desirable solutions and lack of clear authority and process among existing institutions and agencies to implement relocation
>
> ---
>
> 1 http://www.akclimatechange.us/index.cfm, accessed October 8, 2010.

implementation of these plans depends considerably on the complexity of the climate change issues addressed and the capacity to adapt institutional structures to address novel climate change challenges, as discussed in Chapter 5.

Addressing Multiple Interacting Stresses and Time Scales of Response

Climate change is a complex phenomenon that interacts with social, economic, physical, and other drivers of change (Chapter 2; Adger et al., 2009). Along the Gulf Coast of the southeastern United States, for example, intense hurricanes and climate-driven rise in sea level have impacted coastal cities such as New Orleans. This has occurred

once a relocation plan has been agreed to. For example, the Yup'ik Eskimo village of Newtok began working toward relocation in 1994 and obtained authorization from the U.S. Congress in 2003 for its preferred relocation site; however, a school, clinic, and airports cannot be built at the new location because the current population (zero) at the proposed site is less than the minimum required by federal and state statutes to authorize the construction of these facilities (Bronen, 2008). In 2009, the Alaska Department of Transportation built the first infrastructure (a barge landing) at the new village site. During this relocation planning effort, funds have *not* been allocated to repair schools, water and sewage facilities, fuel storage facility, barge landings, and other infrastructure damaged by repeated erosion and flooding because of anticipated abandonment of the current village sites. None of the approximately 25 state and federal agencies that have met bimonthly for a year to develop a relocation strategy has a mandate to assist in relocation. The complex regulations that guide the work of these agencies present barriers to their taking effective action in the relocation effort even though the actions required are well defined (Bronen, 2009). Despite these barriers, the community remains unanimous in its determination to relocate rather than disperse to other communities, which would entail loss of cultural integrity of the community.

This experience in Alaska suggests that, even when adaptation is urgently required to protect life and property, the needed action is agreed upon, and initial funding is available, *current institutions may be ill equipped to implement adaption responses.* Instead, current efforts are directed toward continued planning and protection of existing infrastructure until the relocation can be initiated. For people living in high-risk situations, this continual waiting causes substantial frustration and mental stress (Marino, 2009). Climate-induced human migration in and outside of Alaska will likely become an increasingly central issue, if climate change continues to accelerate. Current estimates put the number of potential climate-induced migrants worldwide at 200 million by the year 2050 (IOM, 2008). Addressing this issue requires substantial institutional innovation and capacity to foster adaptation both in and outside of at-risk communities.

on a landscape where water management had reduced sediment delivery to a protective fringe of barrier islands and where development activities had contributed to land subsidence, making the region more vulnerable to coastal storms (NRC, 2006) (Box 3.3). New Orleans is a hub for agricultural exports and oil imports as well as for transcontinental rail traffic, so impacts of Hurricane Katrina extended far beyond the coastal region (CCSP, 2007). These catastrophic impacts reflected a historical legacy of urban development in areas that had become progressively more vulnerable and of decisions in agricultural, energy, water, ecosystem, transportation, and other sectors that did not adequately consider the potential impacts of climate change (Box 3.3) (AGU, 2006; Colten, 2009). Greater adherence to climate-informed regulations such as

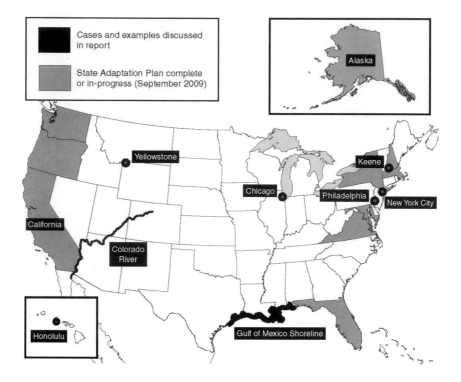

FIGURE 3.1 Locations of case studies described in this report. Within the United States, shading indicates states that have developed climate change adaptation plans as of September 2009. SOURCE: Adapted from Pew Center on Global Climate Change (2009).

building codes, combined with incentives and subsidies, could contribute to climate-appropriate development, phased replacement of infrastructure that spreads the cost of infrastructure adaptation over multiple decades, and relocation of critical transportation corridors away from flood-prone areas (CCSP, 2008e). Similarly, in Germany, recent climate change planning to ensure food security now addresses not only agriculture but also the transportation sector and the global trade network, and several cities and states are integrating climate change adaptation plans into broader smart-growth or sustainability strategies (see the section "Integrating Adaptation Planning into Programs that Address Broader Societal Goals" in this chapter).

Avoiding Maladaptation and Foreclosure of Future Options

One of the most serious risks of a single-issue or sectoral approach to climate change adaptation is that actions taken to alleviate one problem may be maladaptive or foreclose options for future adaptation actions (Adger et al., 2009; Cruce, 2009). This argues for a comprehensive climate change adaptation planning effort at all scales (global, national, state, and local) to provide coordination and to allow for efficient integration of activities. Multisector climate change planning efforts provide opportunities to identify tradeoffs and opportunities and reduce the risk of maladaptive actions. New York City, for example, has involved most sectors of government (e.g., transportation, energy, and environment) as well as stakeholders from the private sector (e.g., utilities), NGOs, and universities in its climate change adaptation planning in an attempt to maximize synergies and minimize unintended tradeoffs (Chapter 4).

In contrast, urban expansion into the wildland-urban interface in the West and fire suppression on adjoining public lands have occurred without adequate consideration of interactions, thereby creating fire risks and vulnerabilities that are maladaptive and create grave risks to life and property (Box 3.4) (Radeloff et al., 2005; Schoennagel et al., 2009). Similarly, flood insurance programs that encourage infrastructure development in floodplains have increased vulnerability to past flooding and to expected increases in flood frequency in the future (see also Chapter 4).

Efforts to focus on climate change actions that address only the immediate issues are particularly likely to create maladaptive long-term responses for at least three reasons (Chapter 4). First, short-term responses often ignore or strongly discount the value of long-term impacts. Second, short-term solutions generally reduce the incentives to explore and develop longer-term solutions that may initially be less cost-effective. Finally, short-term responses to climate change frequently increase the risk to society of longer-term impacts, as in the wildfire example, where fire suppression allows woody fuels to accumulate (Box 3.4), and in the flood insurance example, where incentives encourage development in flood-prone areas (Chapter 4). Similarly, irrigation reduces drought risk in the short term but encourages agricultural development that may exceed the water supply capacity during long-term droughts, as has occurred in arid areas such as Australia, the Colorado River basin, and California (see Chapters 5 and 6). Similarly, subsidies and disaster relief that encourage farmers to rely on drought-sensitive technology or crops or encourage fishermen to intensify fishing effort in response to stock declines reduce incentives for long-term adaptation (Naylor, 2009; Walters and Ahrens, 2009).

In summary, searches for adaptation solutions should consider consequences both for multiple sectors and for the short and the long term. In addition, a comprehensive understanding of the psychological, social, and political obstacles to adaptation is required, as well as an understanding of how to overcome them. Failure to do so frequently increases both vulnerability to climate change and the costs of adaptation

BOX 3.2
Adaptation Challenges Along the Gulf of Mexico

The Gulf Coast, from Galveston, Texas, to Mobile Bay, Alabama, hosts a population of 10 million people spanning four states and is centered on the Mississippi River system and its delta region. Major ports along this coast include Houston, Galveston, Port Arthur, Morgan City, Baton Rouge, New Orleans, Biloxi, and Mobile (AGU, 2006). Two-thirds of U.S. oil imports come through the area, and 90 percent of U.S. domestic outer-continental-shelf oil and gas is transported through the region in pipelines (CCSP, 2008a). Area ports are important for food export from Midwest farms, and the coast has a heavy concentration of chemical manufacturing. The region also holds critical highway and rail links between the eastern and western parts of the nation. This region is at risk from sea level rise and storm surges; adaptation options need to be evaluated to identify appropriate adaptations for both short- and long-term benefits to the region, while also taking national interests into account.

The Gulf Coast population has long been at risk from hurricanes, storm surges, river flooding, global sea level rise, regional subsidence, and a variable hydrologic network (AGU, 2006). Much of the area is rural and poor, with urban poverty as well (CCSP, 2008a). The regional population also scores high on other measures of social vulnerability (e.g., high numbers of disabled persons and elderly), particularly in New Orleans. Many of the risks are mutually reinforcing, and both human activities and climate change seem likely to exacerbate all of these problems. Flood-control measures have reduced the annual flood risk along rivers and, in turn, made lands that were once best left to nature attractive for agriculture and urban growth (NRC, 2006). In the long run, these protection measures have often evolved into maladaptations that contribute to high rates of subsidence, wetland deterioration, and concentration of people and industry in places that are increasingly at risk from storm surges such as those that accompanied Hurricane Katrina (AGU, 2006; NRC, 2006).

The southeastern coastal plain of the United States is low and flat, and land subsidence throughout the region has further contributed to vulnerability to storms and flooding as a result of both natural processes and the extraction of oil and gas, drainage of low-lying areas, and other development activities (AGU, 2006; CCSP, 2008a; NRC, 2006). New Orleans has been subsiding an average of 0.2 inch (5 millimeters) per year. As a result, much of New Orleans now lies below sea level and persists only

over the longer term; it may also reduce incentives to explore more effective long-term solutions.

Managing Adaptively: Learning from Experience

Climate change inevitably creates uncertainty, because future behavior of both the climate system and human decision processes cannot easily be predicted (Chapter 4). Rather than using adaptation strategies to attempt to sustain or preserve a desired past condition, successful adaptation to climate change must attempt to foster con-

because of a system of levees and pumps. In places near Houston, groundwater withdrawal has lowered the terrain by about 6.5 feet (2 meters). Along the Gulf Coast shoreline, such subsidence contributes to a locally high rate of relative sea level rise (AGU, 2006; NRC, 2006). The U.S. Climate Change Science Program's Synthesis and Assessment Product 4.7 (CCSP, 2008a) projects that, by 2050, the Gulf Coast between Mobile and Galveston will see "apparent" sea level rise (actual sea level rise plus land subsidence) of 2 to 4 feet, threatening coastal systems of many kinds: communities, transportation facilities, energy facilities, ecosystems, and others (CCSP, 2008a).

In this case, simply preparing for floods and other risks from extreme weather events offers limited effectiveness as an adaption strategy. Past adaptations to the combined effects of land subsidence and sea level rise through building of levees and other barriers to ocean intrusion and storm surges have proven insufficient to withstand a large hurricane (AGU, 2006; CCSP, 2008a). For example, the Louisiana Department of Transportation (DOT) estimates that storm surges from a strong hurricane (generating wave heights two-thirds the height of those of Katrina) would flood half of the interstate highway system, 98 percent of port facilities, a third of the rail system, and 22 airports (CCSP, 2008a). Given these severe projections, redundancy must be built into critical infrastructure, and a phased relocation of infrastructure and population to less hazard-prone areas warrants serious consideration. If such relocation is built into a smart-growth strategy, the adaptation costs may be modest, given that infrastructure replacement is an ongoing process. Louisiana DOT's current suite of adaptation options are to (1) maintain and manage current infrastructure, absorbing the added maintenance costs; (2) strengthen structures and protect facilities; (3) enhance redundancy; and (4) relocate or avoid hazardous locations (CCSP, 2008a).

Past adaptations to climate variability in the Gulf Coast environment and its resources have provided many benefits; but, as time has passed, many of these actions have constrained present options (AGU, 2006; NRC, 2006). Vulnerable populations, an important national transportation system, and regional ecology are increasingly at risk from decisions made at earlier times. Massive resources continue to be allocated to maintaining the status quo in all these sectors (CCSP, 2008a). Climate change in the form of more frequent or intensive tropical storms, a more intensive precipitation regime and ensuing floods, and accelerated rates of global sea level rise will exacerbate the hazards and make adaptation choices even more difficult (AGU, 2006).

tinued ecosystem integrity and human well-being under uncertain future conditions (Chapin et al., 2009). Thus, neither the target future nor the appropriate methodology for reaching it will be certain over the long term, and society must be prepared to adjust adaptation plans as the future unfolds (Adger et al., 2009). This requires adaptive management, which is defined here as a process of adjusting policies and practices by learning from the outcome of previously used policies and practices (see Chapter 4; Armitage et al., 2007; Chapin et al., 2009; West et al., 2009). Adaptive management is increasingly accepted as best practice in resource management (Armitage et al., 2007). In the Greater Yellowstone Ecosystem, for example, adaptive management provides a

BOX 3.3
Philadelphia

Heat waves can cause significant loss of life, as evidenced in Europe during the summer of 2003 (more than 70,000 excess deaths; Robine et al., 2008) and the 1995 Chicago heat wave (696 excess deaths; Semenza et al., 1996; Whitman et al., 1997). In both cases, the greatest mortality occurred among the elderly and poor.

Heat wave early warning systems can be effective in reducing illnesses and deaths associated with heat waves (Ebi et al., 2004; Palecki et al., 2001; Weisskopf et al., 2002). Partly in response to heat waves in 1993 and 1994, Philadelphia developed its Hot Weather-Health Watch/Warning System (PWWS) in 1995 to alert the city's population when weather conditions pose risks to health (Kalkstein et al., 1996; Mirchandani et al., 1996; Sheridan and Kalkstein, 1998). The city of Philadelphia and other agencies and organizations institute a series of intervention activities whenever the National Weather Service issues a heat wave warning. Television and radio stations and newspapers are asked to publicize the heat wave warning, along with information on how to avoid heat-related illnesses. These media announcements encourage friends, relatives, neighbors, and other volunteers to make daily visits to elderly persons during hot weather to ensure that the most susceptible individuals have sufficient fluids, proper ventilation, and other amenities to cope with the weather. A "Heatline" is operated in conjunction with the Philadelphia Corporation for the Aging to provide information and counseling to the general public on avoidance of heat stress. The Department of Public Health contacts nursing homes and other facilities housing people requiring extra care to inform them of the heat wave warning and to offer advice on the protection of residents. The local utility company and water department halt service suspensions during warning periods. The Fire Department Emergency Medical Service increases staffing in anticipation of increased service demand. The agency for homeless services activates increased daytime outreach activities to assist those on the streets. Senior centers extend their hours of operation of air-conditioned facilities.

An evaluation of the PWWS concluded that the system saved 117 lives during three years of operation (Ebi et al., 2004). The costs of running the system were minor compared with the benefits.

flexible structure for linking climate change projections to a portfolio of actions that address conservation concerns (see Box 3.5). Monitoring the effectiveness of adaptive actions is a key component of adaptive management.

Adaptive management has also been incorporated as a core principle in New York City's planning effort. Likewise, it is a key component of assessing the effectiveness of climate change programs as they are implemented, as in Philadelphia's heat-wave early warning system (Box 3.2), Germany's planning for food security (Chapter 6), and water management programs in Australia and the Colorado River (Chapters 5 and 6).

BOX 3.4
Western Forests and Wildfires

There is mounting evidence that warmer temperatures and drought in the western United States have substantially increased wildfire intensity, areal extent, and frequency (McKenzie et al., 2004; USGCRP, 2009; Westerling et al., 2006). The legacy of fire suppression has increased the buildup of flammable fuels and fire risk in forest types that were previously characterized by low-intensity ground fires (Schoennagel et al., 2004; Swetnam and Baisan, 1995). This increased risk results in part from the overabundance of small-diameter trees that can act as fuel "ladders," enabling flames to reach the forest canopy and cause more damage (Covington and Moore, 1994). The high density of small-diameter trees also makes forest stands, such as Ponderosa pine, more vulnerable to bark beetle infestation, particularly in dry years (Lenart, 2006).

Thinning of small-diameter trees has been used to reduce fire risk in some forests. For example, in the 2002 Rodeo-Chedeski fire in Arizona (Strom and Fule, 2007) and in the Cone fire in California's Lassen National Forest (Hurteau et al., 2008), less damage occurred in stands that had been thinned. Regardless of the forestry benefits, both cost and issues of carbon release to the atmosphere from tree cutting (Hurteau et al., 2008) stand in the way of widespread adoption of forest thinning as an adaptation to climate change. In addition, thinning cannot be viewed as a permanent solution, since forests must be repeatedly thinned to deal with regeneration.

Building of homes on private forest lands (within what is called "the wildland-urban interface") greatly increases risks to life and property from the increased fire frequency and extent in western forests (Radeloff et al., 2005; Theobald and Romme, 2007). It also greatly increases the necessity and cost of fire suppression and eliminates the potential use of prescribed fire as a tool to reduce future fire risks (Schoennagel et al., 2009). A realignment of incentives to discourage private development in fire-prone areas would both reduce the cost and increase the options available for adaptations to climate change.

Adaptive management allows flexibility for climate change programs to adjust when conditions change in unexpected ways, the adaptive plan fails to achieve desired outcomes, or program goals are modified. Adaptive management requires regulatory flexibility for local managers and decision makers to monitor outcomes and adjust appropriately (Armitage et al., 2007; West et al., 2009).

As noted earlier, climate change presents significant challenges for maintaining the viability of species and ecosystems (IPCC, 2007a). The ecological literature contains many general recommendations to safeguard the long-term viability of species and ecosystems, such as increasing connectivity in the landscape, reducing other stressors such as pollution and habitat fragmentation, monitoring to detect changes, intensively managing populations, moving species, and increasing the number of reserves (e.g., CCSP, 2008b; Hannah and Hansen, 2005; Hansen et al., 2003; Heller and Zavaleta,

BOX 3.5
A Conservation Framework for the Greater Yellowstone Ecosystem

To address conservation needs, a working group of academic researchers, scientists from NGOs, and federal resource management agencies is developing a participatory framework that incorporates climate change into natural resource conservation and management by identifying adaptation strategies for species and ecosystems. This framework is being piloted in the Greater Yellowstone Ecosystem (GYE) of Montana, Wyoming, and Idaho (Keiter and Boyce, 1991) to address two conservation concerns: Yellowstone River flows and the grizzly bear (*Ursus arctos horribilis*).

The framework is designed for collaborative application in a given landscape by a multidisciplinary group of experts, including natural resource managers, conservation practitioners, and scientists. The framework draws on expert knowledge to translate climate change projections into a portfolio of adaptation actions. The identified portfolio of actions can then be evaluated in the social, political, regulatory, and economic contexts that motivate and constrain management goals and policies. Application of the framework involves several steps, taken iteratively in an adaptive management context (Armitage et al., 2007):

- Identify conservation targets and specify explicit, measurable management objectives for all targets.
- Build a conceptual model that illustrates the climatic, ecological, social, and economic driver of each target and examine how the target may be affected by multiple plausible climate change scenarios.
- Identify intervention points and potential actions required to achieve management objectives for each target under each climate scenario.
- Evaluate potential actions for feasibility and tradeoffs.
- Prioritize and implement actions.
- Monitor the efficacy of actions and status of management objectives; reevaluate to address system changes or ineffective actions.

The framework was first implemented in 2008 to identify climate-relevant intervention points and potential actions to address management objectives under the initial climate change scenario. For the Yellowstone River, interventions might include manipulation of urban and agricultural withdrawals, stream engineering, and high-elevation check dams. Upland interventions might include increased beaver populations, snowpack management, and forestry activities that influence vegetation structure and wildfire regimes. Potential intervention points for the GYE grizzly bear involve addressing human-bear conflicts (e.g., availability of bear attractants and frequency of human-bear contacts), food resources amenable to management, and connectivity among core habitat areas. Implementation of priority adaptation actions will depend on support within the GYE for investing in novel management approaches.

2009; Lawler et al., 2009; Millar et al., 2007; Scott and Lemieux, 2005) (Table 3.2). Conservation practitioners and agency resource managers have expressed a need for tools to transform this growing menu of recommendations into feasible site- and target-specific strategies for action.

Integrating Adaptation Planning into Programs that Address Broader Societal Goals

Effective adaptation to climate change is only one of many important societal goals. Climate change adaptation is likely to be most effective when it is integrated with ongoing efforts to address other goals rather than when established as a stand-alone program. In New York City and Australia, for example, climate change adaptation was identified as an important new mandate of multiple existing departments and agencies rather than a new stand-alone entity that competed for funds with existing entities (Chapters 4 and 5). Similarly, in Maryland, planning to minimize infrastructure exposure to rising sea level and storm frequency was incorporated into a Smart-Growth policy that protected coastal areas for multiple ecological and social benefits. At the local scale, Keene, New Hampshire, identified many climate change vulnerabilities and developed an adaptation plan that is part of broader planning efforts for sustainability (Chapter 5). NGOs knowledgeable about other planning efforts and state and federal programs whose mandates were consistent with local climate change planning informed and facilitated the development of climate change adaptation plans in Keene. This experience demonstrates the value of coordination and collaboration among federal, NGO, and local entities in planning for climate change adaptation. Along the same lines, in Alaska, a climate change adaptation plan that was initiated to address individual sectors evolved into a more integrated plan because common needs emerged within each sector (e.g., need for downscaled climate projections) and actions taken within each sector depended upon and affected other sectors. Overarching adaptation elements that emerged included a knowledge network to facilitate information exchange between users (agencies, NGOs, businesses, and communities) and scientists and a climate change information and education program.[2]

CONCLUSIONS

In summary, adaptation has emerged as a pressing concern at all levels of government; but action is still hampered by a lack of directed research, poor understanding

---

[2] *http://climatechange.alaska.gov/aag/aag.htm*, accessed October 8, 2010.

of vulnerability, and limited experience with the implementation of adaptation options. The early actions that can be deployed most easily in such an environment are low-cost strategies with win-win outcomes, actions that end or reverse maladapted policies and practices, and measures that avoid prematurely narrowing future adaptation options. Contrary to initial expectations, however, the nation's vulnerability to climate change is very likely to require a substantial effort to adapt appropriately. Recent developments—emergence of scientific consensus on the causes of climate change, evidence that climate change impacts are already under way, a sense that more inevitable changes lie ahead, and a recognition that the past will not be a reliable guide for the future—have validated the need for adaptation planning. They also illustrate that coordination across sectors and across jurisdictional boundaries will be required.

Based on evaluation of current knowledge on adapting to climate variability and recent experience related to climate change adaptation, the panel finds many possible options that are worth considering in responding to observed impacts of climate change and impacts projected in the relatively near future. At the same time, benefits, costs, potentials, and limits of these options for adapting to climate change impacts are generally not well understood.

Adaptation planning is under way in many states and localities in the United States and by some nongovernmental groups concerned about sustainability, resilience, and changing market conditions. Much of this planning is oriented toward protecting current systems and infrastructures and maintaining the status quo (e.g., Gulf Coast, Alaska). However, maintaining the status quo may in many cases be maladaptive over the long term if it draws attention and resources away from implementation of novel longer-term solutions.

Many climate change adaptation plans emphasize *either* societal impacts (e.g., Gulf Coast) *or* ecological impacts (e.g., Greater Yellowstone Ecosystem). Plans that integrate the twin goals of societal well-being and ecological integrity emerge in places (such as New York City) where both social and environmental goals have been clearly articulated in an integrated planning mission statement. Climate change adaptation plans and actions are most likely to meet society's needs when they remove incentives for maladaptive responses (e.g., inadequate steps with fire suppression, fisheries subsidies) and avoid foreclosing future options. Risks of maladaptation can be reduced by integrating efforts to adapt and to limit the magnitude of climate change in a common sustainability agenda. In addition, partnerships that involve federal, state, and local agencies and also engage NGOs, utilities, and private businesses are particularly effective in developing comprehensive plans for climate change adaptation. Such an integrated approach reduces the likelihood that each agency and stakeholder group will pursue a separate and partially incompatible agenda.

Climate change adaptation plans are particularly relevant to local populations when they incorporate programs, plans, and experience that address *current* climate extremes such as severe storms, floods, droughts, and heat waves (e.g., New York City, Philadelphia, United Kingdom). These and other extreme events are projected to become more frequent in the future. Adaptation options are much more limited to cope with impacts of relatively severe climate change in the longer run, if efforts to limit emissions are not successful (Chapter 2). Some projected impacts are likely to be beyond the scope of adaptation unless it involves major structural change. An important part of a national approach to adaptation is looking toward the potential for these more severe impacts and considering possible limits to adaptation.

Conclusion: Searches for adaptation solutions must consider consequences both for multiple sectors and for the short and long terms.

Conclusion: In the short term, it is advisable to use familiar options that are likely to be effective in addressing relatively near-term needs for adaptation. Limited knowledge about future impacts is not an excuse for inaction.

Conclusion: Current adaptation options, however, are supported by a limited body of experience and evidence, which means that adaptation practice in the United States needs to be coupled with a strong effort not only to improve what we know about options already under discussion but in some cases to learn about and devise novel approaches not currently under discussion.

Conclusion: Appropriate short-term climate change adaptation options should be strengthened and sustained by those entities best poised to implement them. This often requires development of minimal federal standards to ensure a coordinated national effort, federal funding or financing to provide incentives for state and local actions, and state and/or local implementation to ensure that adaptations make sense in the local context (Chapter 5).

Conclusion: Policies and/or institutions should be modified or established that align incentives for the private sector to engage more effectively in short- and long-term climate adaptation solutions.

Conclusion: Appropriately funded actions should be taken to support effective long-term geographic, sectoral, and social adaptation to climate change *before* rather than after disaster strikes, both in the near term and in the longer term.

TABLE 3.2 Ecosystems: Examples of ideas about specific options for facilitating ecosystem adaptation to climate change and identification of entities best poised to implement each option.

Climate Change	Impact	Possible Adaptation Action[a]	Federal	State	Local Government	Private Sector	NGO/Individuals
Altered hydrologic regime	Water stress on ecosystems	Manage for high water-use efficiency and drought-tolerant species in areas of increasing drought (4-4; CCAL; DOI-LW; NEC).		■	■	■	■
		Establish guidelines to protect against stream drying; establish watershed-monitoring programs; recharge groundwater when availability exceeds requirements for ecosystems and society.	■	■	■		
		Purchase or lease water rights and wetlands to provide species refuges and enhance flow management options (4-4; CCAL).	■	■	■	■	■
		Backfill canals and mosquito ditching to prevent saltwater intrusion (DOI-LW).			■		■
	Flow effects on rivers	Plant flood-adapted species to reduce peak flows and erosion. Develop more effective stormwater infrastructure to reduce severe erosion (4-4).			■		■
		Manage reservoir releases to provide cold water downstream (NEC; WWF). Reforest riparian areas with native species to create shaded thermal refuges for fish species if water supply is adequate (4-4; NEC).	■	■	■		■
Climate change generally	Degradation of ecosystems	Reduce atmospheric CO$_2$ concentrations to limit ocean acidification	■	■	■	■	■
		Use buffers as barriers and time construction and maintenance activities to avoid favoring spread of invasive species (WWF). Use desirable nonnative species to compete with undesirable invasives when native species cannot (DOI-LW).			■	■	

	1	2	3	4	5
Enhance resilience by managing for diversity of species, genotypes, and habitat heterogeneity where climate trends are uncertain: establish special protection for areas that support keystone processes or sensitive species (4-4; DOI-LW; NEC; WWF).	■	■	■		
Restore and increase habitat availability, and reduce stressors, in order to capture the full geographical, geophysical, and ecological ranges of species on as many refuges as possible (4-4).	■			■	
Realign management targets in the context of climate-induced changes, rather than seeking to maintain or restore a "reference" condition that is no longer climatically viable (4-4; CCAL; DOI-LW).					■
Protect areas along climate gradients to provide a wide range of climate adaptation options (WWF); protect areas that have enduring features unlikely to be affected by climate change (e.g., high relief, limestone karst) (WWF).					■
Identify climate change refuges, assess the optimal size, and acquire the necessary land (4-4; CCAL; OIGCC). Use an insurance factor when calculating reserve sizes to account for uncertainty in climate change (WWF).	■			■	■
Minimize management of some areas (e.g., wilderness) and protect migration corridors to allow colonization and successional processes to adjust naturally (4-4; WWF).	■		■	■	
Proactively manage early successional stages that follow widespread climate-related mortality by promoting diverse age classes, species mixes, stand diversities, genetic diversity, etc., at landscape scales (4-4; NEC).	■	■		■	
Facilitate natural adaptation by management practices that shorten regeneration time and foster interspecific competition (to speed rate of species change) (4-4).	■	■		■	
Use paleoecological records and historical ecological studies to revise and update restoration goals so that selected species will be tolerant of anticipated climate and/or habitats will be buffered from climate change (e.g., altitudinal gradients) (4-4; NEC; CCAL; WWF)	■	■		■	

Category	Subcategory	Strategy					
		Plant appropriate native (or, if necessary, introduced) desired species after disturbances or in anticipation of the loss of some species (4-4).	■	■	■		
		Assisted migration where appropriate: explore the establishment and growth of plant species more adapted to expected future climate conditions (4-4; CCAL; DOI-LW). Test the suitability of species that are nonnative locally, but sustain native biodiversity or enhance ecosystem function regionally (4-4). Rather than maintaining only historic distributions, spread species over a range of environments according to modeled future conditions (4-4; CCAL).	■	■	■	■	■
Climate change generally	Problems for native species	Identify and take early proactive action against nonnative invasive species that respond to climate change, especially where they threaten native species or current ecosystem function (4-4; NEC; CCAL; OIGCC; WWF; DOI-LW).	■	■	■		
		Use climate change data to refine threatened or endangered status (DOI-LW).					■
		Reduce or eliminate stressors or harvest on conservation target species (4-4; WWF). Establish no-take hunting and fishing areas for climatically threatened species (WWF). Create or protect refuges for valued aquatic species at risk to the effects of early snowmelt on river flow (4-4). Provide redundant refuge types to reduce risk to trust species (4-4; NEC; WWF). Practice bet-hedging by replicating populations and gene pools of desired species (4-4; WWF).				■	■
		Use conservation easements and buffers around refuges to foster population and species variability and to provide room for species dispersal and landscape interactions (4-4; WWF; DOI-LW).	■		■	■	
		Establish or strengthen long-term seed banks and preserve species in zoos and botanical gardens to create the option of reestablishing extirpated populations in new or more appropriate locations (4-4; DOI-LW; WWF). Facilitate interim propagation and sheltering or feeding of mistimed migrants, holding them until suitable habitat becomes available (4-4).	■	■			

More extreme events		Assess tradeoffs of long-distance transport or assisted migration of threatened and endangered endemic species (4-4); modify genetic diversity guidelines to increase the range of species, maintain high effective population sizes, and favor genotypes known for broad tolerance ranges (4-4); remove dispersal barriers, including dams, establish dispersal bridges, and connect landscapes, that support migration of native species in response to climate change (4-4; NEC; CCAL; OIGCC; WWF); reintroduce lost native species (WWF); identify species or habitats that are likely to migrate out of areas established for their protection (DOI-LW).	■	■	■
Increased wildfire		Allow natural fires to burn where they sustain long-term ecosystem integrity; restore natural disturbance regimes (NEC; WWF); include climate change in National Fire Plan to effectively achieve conservation goals (NEC).	■	■	
		Facilitate climate-appropriate fire regime through fuels management, wildland and prescribed fire, and suppression (4-4; CCAL; DOI-LW); minimize alteration of natural disturbance regimes, for example through removal of infrastructure that prohibits the allowance of wildland fire or changes in incentives to discourage building in the wildland-urban interface (4-4; NEC; WWF).	■	■	■
		Proactively manage early successional stages that follow widespread climate-related mortality by promoting diverse age classes, species mixes, stand diversities, genetic diversity, etc., at landscape scales (4-4; NEC).	■		
Climate change generally	**Threats to ecosystem services**	Multiple-use management to maintain cultural and aesthetic services, as well as recreational and tourism potential.	■	■	■
		Participate in carbon markets that will lead to forest preservation or restoration (DOI-LW).		■	■
		Manage for agricultural and forestry products, ecological integrity, and livelihoods.		■	■
		Conserve and manage lands suitable for carbon sequestration and other climate feedbacks (CCAL; DOI-LW).	■	■	

Outdated management	Incorporate climate change effects into Environmental Protection Agency (EPA) water quality, air quality, and groundwater standards.	■					
	Identify resources and processes most sensitive to climate change through monitoring and assessment programs (4-4; NEC; CCAL; DOI-LW) and scenarios that explore the societal consequences (4-4). Incorporate long-term monitoring into design and management changes to ensure they are responsive to changes in base conditions.		■	■	■		
	Update legislation to require anticipatory (rather than historical) guidelines and reference points, and use longer planning horizons (NEC; CCAL).	■	■				
	Rapidly assess (e.g., using the Resource Planning Assessment Process) existing federal agency plans, contract language, and leases to determine the level of preparedness for climate change risks (4-4; DOI-LW).	■					
	Form partnerships among federal and state agencies, the private sector, local people, and other stakeholders to address climate change and its interactions with landscape-scale human-caused stressors (4-4; DOI-LW; CCAL; WWF).	■	■	■	■	■	
	Remove structures that harden the coastlines, impede natural regeneration of sediments, and prevent natural inland migration of sand and vegetation in response to climate change (4-4); restore or create coastal wetlands, barrier islands, and other protective natural ecosystems.		■	■			

NOTE: Most adaptations are local and need to be tailored to local conditions. The suitability of each adaptation option must therefore be evaluated in the context of local conditions. Where possible, the table refers to assessments and syntheses that consider multiple adaptation options and provide references to specific studies.

[a] Creation of novel ecosystems warrants extreme caution, given the past checkered history of species introductions. This approach might be appropriate where a current ecosystem is substantially degraded by climate change or other human caused impacts.

SOURCE: Reference citations were abbreviated as follows to conserve space: NEC (Glick et al., 2009), CCAL (Theoharides et al., 2009), 4-4 (CCSP, 2008b), OIGCC (Parmesan and Galbraith, 2004), WWF (WWF, 2003), DOI-LW (DOI, 2008b).

TABLE 3.3 Agriculture and Forestry: Examples of ideas about specific options for facilitating agriculture and forestry sector adaptations to climate change and identification of entities best poised to implement each option.

Climate Change	Impact	Possible Adaptation Action	Federal	State	Local Government	Private Sector	NGO/Individuals
Climate change generally	Need for more intense management	Use remote sensing to monitor broad-scale spatial patterns, like shifts in plant community composition, vegetation production, changes in plant mortality, rise of invasive species, deforestation, etc. (4-2) develop cost-effective strategies to address climate change through integration of models based on long-term monitoring of agricultural lands (4-3).	■	■			■
	Potential new markets and new competition; need to adapt to mitigation measures	Respond to market signals and conduct research to anticipate global market trends (NRC; IPCC); diversify farm operations (IPCC); plan for changes in fuel and fertilizer expenses; develop new seed varieties, technologies, and practices; anticipate consumer demand for local crops without excess embedded carbon.				■	■
	Impacts to ecosystem services	Preserve biological diversity as a natural buffer against climatic shocks (NRC); move to no till agriculture to increase the carbon captured in the soils.	■	■			
	Loss of crop yield	Foster diversification and innovation by modifying subsidy, support, and incentive programs.	■				

		Use technologies to "harvest" water, conserve soil moisture, and to use water more effectively in areas with rainfall decreases (IPCC); adopt irrigation best practices (e.g., drip irrigation, laser leveling, etc.); switch to crops with greater drought resistance; cease irrigated agriculture (or cap water withdrawals) in regions where groundwater is being depleted; reduce stocking density in forests to offer resilience to drought, insects, and fires.		■			■
Increased evaporation and changes in precipitation; increased precipitation	Greater irrigation requirements	Manage to prevent waterlogging, erosion, and nutrient leaching in areas with rainfall increases (IPCC); develop flood-resistant crops.	■	■		■	■
	Increased yields of rain-fed agriculture				■		■
	Impacts to tree viability	Reforest with genetically diverse seed sources (Millar); move forestry operations to regions projected to have more suitable future climate; manage soil and groundwater to prevent waterlogging.				■	■
Sea level rise	Brackish water infiltrating coastal farmland (4-1)	Backfill irrigation ditches; install tide gates to control flooding.	■	■	■	■	■
Increased temperature: effects on pests	Greater spread of *animal diseases* from low to middle latitudes due to warming (IPCC)	Change breeding practices; move to new lands for cattle grazing.	■				■
	Climate change will likely lead to a northern migration of *weeds* (4-3)	Improve the effectiveness of pest, disease, and weed management practices through wider use of integrated pest and pathogen management that forecast potential new pest incursions (driven by climate change), assess tools that are currently available to combat these pests (e.g., eradication, containment, management, and existing pesticides), and what may need to be developed if gaps currently exist (IPCC); diversify species stocking and reduce stocking density in planted forests.	■	■	■	■	■

Category	Impact	Adaptation
	Disease pressure on crops and livestock will increase with earlier springs and warmer winters, allowing low proliferation and higher survival rates of *pathogens and parasites* (4-3)	Develop and use disease resistant varieties.
Increased temperature: effects on livestock	Climate changed-induced shifts in *plant productivity* and type will impact livestock operations (4-3)	Shift grazing to new lands; shift to species and breeds more resistant to droughts and reduce animal stockir g density; change feedstocks.
	Higher temperature will likely reduce livestock production during the summer; partially offset by warmer winters (4-3; IPCC)	Provide more shade opportunities, improve air flow in barns, use air conditioning, and change breeds; supplement feed during periods of low forage availability.
	Increased temperature will *lengthen growing season*, extending forage production season and decreasing the need for winter season forage reserves (4-3)	Adjust forage reserves.

Increased temperature: effects on crops	Crops near climate thresholds, like grapes, may decrease in yield and/or quality (IPCC)	Improve seeds (IPCC) and maintain and diversify strains of crop varieties to the climate of the current decade (NRC); use climate scenarios to identify areas where agriculture will become more favorable; shift lumber production from areas of decreasing favorability to those of increasing favorability (NRC); reforest with species genetically adapted or compatible with the anticipated future climate (NRC); find alternatives for timber (NRC).
	Improve the climate for fruit production in Great Lakes region and eastern Canada but with risks of early season frost and damaging winter thaws (IPCC)	Develop varieties that can withstand early frost.
	High temperatures during flowering may *reduce grain number, size, and quality* (IPCC); *pollen sterilization by* extreme temperatures (NRC)	Develop hybrids that can tolerate higher temperatures.
	Possible changes in the *length of the growing season* and in growth rates changing the required growing season (NRC)	Change planting dates and cultivars to respond to changed climate (NRC; IPCC); shorten the rotation of managed forests to be more responsive to climate (NRC).

	Effect	Adaptation strategy					
	Possible changes in the sensitivity of plants to *fertilizers, pesticides, and herbicides* (NRC)	Invest in improved varieties of trees, forest protection, forest regeneration, silvicultural management, and forest operations (IPCC); consider other values of forests (e.g., watershed management, recreation) where commercial forestry is no longer climatically suitable.	■	■	■	■	■
Higher atmospheric CO_2 concentrations	Increased production of some trees and crops	Consider new varieties: fast-growing trees may respond strongly to increased CO_2 and increase productivity (IPCC); increased CO_2 and temperature will speed life cycle of grain and oilseed crops (4-3).	■				
	Stresses will accumulate over time	Switch to annual crops instead of perennials (IPCC).	■	■			
	Selective advantages for C3 weeds: greater competition with crops	Select crops and cropping practices that reduce competitive success of weeds, including the adjustment of integrated pest management approaches to new conditions.	■	■			
Extreme events	Increased climate extremes may promote *plant disease and pest outbreaks* (IPCC)	Produce hybrids more resistant to the diseases and pests.	■	■			
	Increased risk of *flood* in some regions (4-2)	Change crop insurance processes; change lands used to minimize flood risk; protect wetlands buffers.	■				■
	More frequent extreme events may lower yields by *directly damaging crops* at specific developmental stages, like flowering, or making the timing of field applications more difficult (IPCC)	Breed or genetically modify crop varieties to have greater tolerance of heat, drought, and flood extremes.	■	■			

	Impact	Adaptation option						
	Decrease in snow cover and more winter rain on bare soil lengthen erosion season and *enhance erosion* (IPCC)	Change cropping areas; adopt reduced tillage and other best practices to mitigate erosion.	■		■			
	Forest fires will become more common as climate becomes hotter and drier (NRC)	Lower stocking density to reduce fire regime type from lethal to mixed or sublethal; thin small-diameter trees; increase use of prescribed fire in dry forest areas to reduce likelihood of intense, lethal, natural fires; restructure carbon storage calculations that unduly penalize forest thinning for fire management; shift forest production to less fire prone regions.			■	■	■	
	Point and non-point source pollution from agriculture practices could increase	Use buffers, adjust fertilizer and pesticide applications, and adopt other best practices to limit pollution impacts on water resources.	■	■				

NOTE: Most adaptations are local and need to be tailored to local conditions. The suitability of each adaptation option must therefore be evaluated in the context of local conditions. Where possible, the table refers to assessments and syntheses that consider multiple adaptation options and provide references to specific studies.

SOURCE: Reference citations were abbreviated as follows to conserve space: 4-1 (CCSP, 2009b) 4-2 (CCSP, 2009a), 4-3 (CCSP, 2008c), NRC (NRC, 1992), IPCC (IPCC, 2007a), Millar (Millar et al., 2007).

TABLE 3.4 Water: Examples of ideas about specific options for facilitating water sector adaptation to climate change and identification of entities best poised to implement each option.

Climate Change	Impact	Possible Adaptation Action	Federal	State	Local Government	Private Sector	NGO/Individuals
Higher temperature and reduced precipitation	Insufficient water supplies	Enhance supplies through traditional supply approaches including dams, larger reservoirs and other storage facilities, importing water, or transferring water between basins (IPCC4; IPCC3; CALI; NRC). Other approaches include increasing system redundancy to ensure backup supplies, sharing integrated facilities between jurisdictions and sectors, obtaining a portfolio of multiple sources of water, including reuse of municipal wastewater (IPCC4; IPCC3; USGS; NRC; CCAWS).	■	■	■		
		Purchase alternative supplies through water trading and exchange (USGS; IPCC4); store water during wet years or seasons (conjunctive management).	■	■	■	■	
		Participate in water supply protection through watershed management, including protecting surface water sources and groundwater recharge zones.	■	■	■	■	■
		Encourage water harvesting and gray water use (NRC; IPCC4; CALI; IPCC3); design sites to minimize water requirements (e.g., low-water-use landscaping) and retain gray water and stormwater on site for landscape purposes (NRC; CALI; IPCC4).	■	■	■	■	■
		Regulate water use more stringently, restrict specific uses of water, and adopt best practices for conservation and demand management in all sectors (CALI; CUWCC; IPCC3; IPCC4; NRC; USGS; CCAWS).	■	■	■	■	■

Category	Strategy					
	Consider reform of water allocation by allocating a percentage of available supplies rather than a fixed volume, establishing a water rights entitlement for the environment, downsizing or abandoning parts of a system, updating monitoring and accounting of water rights systems, enacting market reforms to allow interstate trading, and compensating rights holders and assisting in transition (FPB).	■	■	■	■	■
	Design pricing policies to encourage water conservation and to respond to drought or long-term storage conditions (CCAWS).		■	■	■	■
Inadequate water for ecosystems	Use water banking and other market mechanisms to augment supplies, regulatory or incentive programs to protect or enhance instream flows to support habitat, environmental mitigation programs to offset damage caused by new projects, contracts to access water during dry years, etc., to ensure supply (USGS; IPCC4).	■	■	■	■	■
	Revise or update environmental regulations to facilitate resolution of competing demands for water in light of changing conditions (e.g., adaptive management).			■	■	■
	Purchase water rights for environmental protection (5-1).	■	■	■	■	■
Decreased snowpack in West and Northeast	Enhance reservoir storage and aquifer storage capacity, reoperation of reservoirs, water transfers, and vegetation management to enhance water storage and manage timing of runoff from watersheds.			■	■	■
More variable streamflow and lake levels	Build or reoperate dams to store water and control downstream releases (taking into account the likely environmental consequences).			■	■	■
Changed seasonality of precipitation	Change dam and storage facility operations to offset timing issues; reduce diversions and find alternative supplies during low flow periods (CALI).		■	■	■	■

		Strategy						
Storm surges, sea level rise, and increasing intensity of precipitation	Increased frequency of coastal and riverine flooding	Update FEMA floodplain maps to reflect higher probability of high impact events.			■			
		Evaluate risks to infrastructure and develop and apply new design standards for water, wastewater, and drainage systems (USGS).		■	■	■		
		Enhance regulation of floodplain development; change design standards to allow floods to pass under buildings (e.g., pilings); encourage relocation of infrastructure from areas where flooding and erosion are likely and retreat from damaged areas after flooding, especially in 100-year floodplain (USGS; IPCC3).		■	■	■	■	
		Use climate forecasts to optimize reservoir operations and flood control storage.			■	■		
		Redesign flood-prone areas to allow natural attenuation processes, reduce hard surfaces, allow natural channel movement, etc.			■	■		
		Design new or improved levees, dikes, and flood walls to withstand higher flood levels (IPCC3); enhance dam safety inspections and modeling and consider relocation where engineering solutions make life and property more vulnerable to extreme events (CCAWS).		■	■	■		
		Protect vulnerable land from development through land use planning.			■	■		
	Increased levels of pollutants in runoff	Enhance flood retention and buffer requirements, natural filtration capacity, and biological removal of pollutants; implement rain gardens (CALI).	■	■	■	■		
		Increase funding for water quality regulation and remediation.		■	■	■		
	Increased stormwater runoff	Require treatment of urban stormwater runoff, manage land uses to require onsite retention in areas where pollutants are generated.		■	■	■		

Category	Effect	Adaptation strategy					
Higher water temperature	Ecological effects	Change dam operation to release more cold water (CALI; WWF).			■	■	■
		Raise dissolved oxygen levels with aeration devices or re-oxygenation.			■	■	
	Increased organic material in public water supplies	Adjust seasons for recreational fishing, or fish at new locations.	■		■	■	■
		Protect upstream watersheds and increase monitoring of water quality (CALI; IPCC3).			■	■	
		Consider new disinfection standards or alternative disinfection approaches; adopt carbon filters to improve drinking water quality.	■		■	■	
	More stratification in lakes	Change dam operations to minimize effects on water chemistry and biology (WWF).			■	■	
Sea level rise	Saline intrusion into coastal aquifers	Insert sea water barrier injection wells (to limit migration of saltwater aquifers inland), e.g., with reclaimed water (CALI).			■	■	
		Desalinate (IPCC3; IPCC4; USGS; NRC).		■	■	■	
		Reduce pumping of coastal aquifers, move wells further from the coast.			■	■	
	Saline intrusion into estuaries, inundation of coastal wetlands	Create physical barriers to prevent ocean water from entering estuaries and wetlands; allow for inland migration of wetlands.		■	■	■	
		Reduce diversions from coastal rivers (CALI).			■	■	■
General climate change	Landscape impacts on water supply	Manage forests to reduce large-scale fire hazard as well as erosion and sedimentation after forest fires (CALI).		■	■	■	
		Regulate landscaping and building materials in urban-wildland interface.			■	■	■
		Protect important water-based habitat from invasive species (IPCC4; NRC; CALI); identify, restore, and protect important ecosystems, especially those with endemic species that are at high risk (WWF).	■	■	■	■	■

Outdated institutions in light of changing conditions	Adaptation option					
	Encourage flexibility in water management rules, and move away from using strict formulas based on assumptions of stable, stationary climate (5-1; IPCC4; CALI; IPCC3).			■	■	■
	Develop innovative tools and methods to manage water resources in light of change and uncertainty (5-1; IPCC4; USGS; IPCC3).	■	■	■	■	■
	Update professional training and university curricula to incorporate a nonstationary climate and uncertainty (5-1; CCAWS).	■	■	■	■	
	Encourage collaborative regional water supply planning to address multiple stresses including climate change.			■	■	
	Maintain current networks for monitoring of snowpack, flows, and other conditions (CALI; 5-1).				■	■
	Improve use of monitoring data; develop tools to better incorporate recent trends in management processes (adaptive management) (CALI; 5-1).	■	■	■	■	■

NOTE: Most adaptations are local and need to be tailored to local conditions. The suitability of each adaptation option must therefore be evaluated in the context of local conditions. Where possible, the table refers to assessments and syntheses that consider multiple adaptation options and provide references to specific studies.

SOURCE: Reference citations were abbreviated as follows to conserve space: 5-1 (CCSP, 2008d), IPCC3 (IPCC, 2001b), IPCC4 (IPCC, 2007a), CALI (California Department of Water Resources, 2008), NRC (NRC, 2007b), USGS (Brekke et al., 2009), CCAWS (Ludwig et al., 2009), CUWCC (California Urban Water Conservation Council, 2008), FPB (Young and McColl, 2008), WWF (WWF, 2003).

TABLE 3.5 Health: Examples of ideas about specific options for facilitating health sector adaptation to climate change and identification of entities best poised to implement each option.

Climate Change	Impact	Possible Adaptation Action	Federal	State	Local Government	Private/NGO	Individuals
Changes in the frequency, intensity, and duration of extreme weather events (i.e. floods, droughts, windstorms, wildfires)	Increased risk of injuries, illnesses, and death	Develop scientific and technical guidance and decision-support tools for early warning systems and emergency response plans, including appropriate individual behavior (Ebi).	■	■	■	■	
		Implement early warning systems and emergency response plans, including medical services (J and S; Ebi; Frumkin).	■	■	■	■	
		Conduct education and outreach on emergency preparedness and response, including mental health needs following a disaster (Ebi; Frumkin).	■	■	■	■	
		Conduct tests of early warning systems and response plans before events (Ebi).	■	■	■	■	
		Monitor and evaluate the effectiveness of systems.	■	■	■	■	
		Provide scientific and technical guidance for building and infrastructure standards to reduce hazards (Ebi; Frumkin).	■	■	■	■	
		Develop and enforce building and infrastructure standards that take climate change into account and reduce hazards (Ebi; Frumkin).	■	■	■	■	
		Monitor the air, water, and soil for hazardous exposures following floods, windstorms, and wildfires (J and S; Ebi).	■	■	■	■	
		Stay informed about impending weather events (Ebi).		■	■	■	■
		Follow guidance for emergency preparedness, and for conduct during and after an extreme weather event (Ebi).					■

Stressor	Health risk	Adaptation strategy					
Increases in the frequency, intensity, and duration of heat waves	Increased risk of heat-related illnesses and deaths	Develop scientific and technical guidance and decision-support tools for heat wave early warning systems and emergency response plans, including appropriate individual behavior (J and S; Ebi).		■	■	■	■
		Implement heat wave early warning systems and emergency response plans, taking climate change into account (J and S; Ebi; Frumkin).		■	■	■	■
		Conduct education and outreach on preparedness during a heat wave (J and S; Ebi).		■	■	■	■
		Develop education and training programs for health professionals on the risks of and appropriate responses during heat waves (J and S; Ebi).		■	■	■	■
		Monitor and evaluate the effectiveness of heat wave early warning systems (J and S; Ebi).		■	■	■	■
		Improve urban design to reduce urban heat islands by planting trees, increasing green spaces, etc. (Ebi).		■	■	■	■
		Improve building design to reduce heat loads during summer months (Ebi; Frumkin).		■	■	■	■
Warmer temperatures on cloudless days	Increased risks of adverse health outcomes related to poor air quality	Modify and enforce air pollution regulations as necessary to take climate change into account (Ebi; Frumkin).				■	■
		Develop early warning and response systems for days with poor air quality (J and S; Ebi).		■	■	■	■
		Conduct education and outreach on the risks of exposure to air pollutants (J and S; Ebi; Frumkin).		■	■	■	■
		Follow medical advice on appropriate behavior on days with high concentrations of ozone, particulate matter, airborne allergens, and other air pollutants (Ebi).	■				
Changing in mean and extreme temperature and precipitation	Changes in the geographic range and incidence of vector-borne and zoonotic diseases	Develop scientific and technical guidance and decision-support tools for early warning systems (Ebi).		■	■	■	■
		Implement, modify, and sustain early warning systems for vectorborne and zoonotic diseases to take climate change into account (J and S; Ebi).		■	■	■	■

		■	■	■	■
Modify vector (and pathogen) surveillance to take climate change into account (Ebi).		■	■	■	■
Disseminate information on appropriate individual behavior to avoid exposure to vectors, including eliminating vector breeding sites around residences (J and S; Ebi).		■	■	■	■
Disseminate information on signs and symptoms of disease to guide individuals on when to seek treatment (J and S; Ebi).		■	■	■	■
Provide low-cost vaccinations to those likely to be exposed (Ebi).		■	■	■	■
Sponsor research and development on vaccines and other preventive measures (Ebi).	■	■		■	■
Consider possible impacts of infrastructure development, such as water storage tanks, on vector-borne diseases (Ebi).			■	■	■
Changes in the geographic range and incidence of waterborne and foodborne diseases — Provide scientific and technical guidance and decision-support tools for early warning systems (Ebi).		■	■	■	■
Modify and enforce watershed protection regulations to take climate change into account (Ebi; Frumkin).			■	■	■
Modify and enforce safe water and food handling regulations to take climate change into account (Ebi).			■	■	■
Modify water and wastewater treatment facilities, and drainage and stormwater management to take climate change into account.			■	■	■
Evaluate consequences of placement of sources for possible contamination by water- and foodborne pathogens (Ebi).		■	■	■	■
Follow guidelines on drinking water from outdoor sources (Ebi).	■	■	■	■	■
Disseminate information on signs and symptoms of disease to guide individuals on when to seek treatment (J and S; Ebi).	■	■	■	■	■

Climate change generally	Institutional challenges		Federal	State	Local Government	Private Sector	NGO/Individuals
		Modify public health programs and activities focused on climate-sensitive health outcomes to take climate change into account (J and S).	■	■	■	■	■
		Enhance education of health care professionals to understand the health risks of climate change, including diagnosis and treatment for health outcomes that may become more prevalent (J and S).	■	■	■	■	■
		Provide federal leadership for health organizations and agencies to effectively collaborate and coordinate on research, development of decision-support tools, and other activities (J and S; Frumkin).	■				

NOTE: Most adaptations are local and need to be tailored to local conditions. The suitability of each adaptation option must therefore be evaluated in the context of local conditions. Where possible, the table refers to assessments and syntheses that consider multiple adaptation options and provide references to specific studies.

SOURCE: Reference citations were abbreviated as follows to conserve space: Ebi (Ebi et al., 2008), Frumkin (Frumkin et al., 2008), J and S (Jackson and Shields, 2008).

TABLE 3.6 Transportation: Examples of ideas about specific options for facilitating transportation-sector adaptation to climate change and identification of entities best poised to implement each option.

Climate Change	Impact	Possible Adaptation Action	Federal	State	Local Government	Private Sector	NGO/Individuals
Long-term sea level rise	Permanent flooding of coastal land	Build or enhance levees/dikes for protection.	■	■	■	■	
		Elevate critical infrastructure that is at risk for sea level rise.	■	■	■	■	
		Abandon or move threatened facilities to higher elevations.	■	■	■	■	

Climate change and potential impacts	Adaptation strategy
Loss of barrier islands	Protect and/or relocate newly exposed railroads, highways, and bridges.
	Switch to alternate shipping methods if waterborne transport cannot use the Intracoastal Waterway or other shipping channels.
Impacts on infrastructure such as bridges or harbors (RFF-PI)	Raise bridge heights and reinforce or relocate harbor infrastructure.
New patterns of prevailing winds — Existing airport runways may become less efficient; time of travel on long-distance flights and transoceanic shipping may be affected	Increase airport runway lengths.
Time of travel on long-distance flights and transoceanic shipping may be affected	Evaluate effects on logistics; adjust schedules.
More intense precipitation — Change in hydrology	Revise hydrologic flood frequency models.
Change in hydrology	Revise computational models for storm return frequencies.
Change to hydraulics	Revise design standards for hydraulic structures (culverts, drainage channels, and highway underpasses).

Climate stimulus	Adaptation strategy
More frequent flooding	Reinforce at-risk structures with particular attention on bridge pier scouring.
	Review hydraulic structures for deficiencies (culverts and drainage channels).
	Provide federal incentives to avoid development in flood plains.
	Institute better land use planning for floodplain development including prohibition in some instances.
	Recognize the inherent cost to society of construction in flood-prone areas.
	Elevate structures where possible; reconstruct to higher standards.
	Replace vulnerable bridges and other facilities.
	Harden infrastructure and port facilities.
Changes in efficiency of some transportation modes; change in safety (or perception of safety) in some transportation modes	Shift transportation preferences among air, rail, ship, or highway routing as appropriate.
Stress on pavements and road decks	Research new pavement materials and bridge decking materials that are more resistant to heat stress and degradation.
	Establish standards for and use heat-resistant pavements.
Warmer temperatures and heat waves	Replace vulnerable pavement, outdated expansion joints, or runways as needed.
	Revisit Occupational Safety & Health Administration standards for construction workers in light of higher temperatures and other climate stresses.

Impact	Sub-impact	Strategy							
	Railway buckling	Implement more nighttime construction.			■	■			
		Research on stresses in rails leading to buckling.	■	■	■				■
		Implement changes in rail design to accommodate higher temperatures to prevent rail buckling.			■				
	Great Lakes water level reductions, lower flows in major rivers	Implement changes in shipping vessels or freight weights.		■	■				
		Find alternatives to barges and water transport.	■	■	■				■
		Dredge channels to greater depths.	■		■	■			
	Lower air density	Increase airport runway lengths.	■	■	■	■			
	Longer ice-free periods	Lengthen the shipping season on inland waters and in the Arctic (RFF-PI).		■	■	■			
		Extend shipping to previously inaccessible areas.			■				
	Changes to engine fuel efficiency	Changes (+/-) to the amount of fuel needed in all forms of motorized transport.	■					■	
		Reevaluate airport runway lengths required for take-off.	■	■	■	■			■
Cold regions impacts	Loss of permafrost	Research on pavement design over thawing permafrost.	■						
		Identify areas and infrastructure vulnerable to thawing permafrost.		■	■				
		Revise roads, bridge foundation, runway, and railway design criteria and standards to reflect loss of permafrost.		■	■				
		Replace at-risk roads, runways, and railroads.	■	■	■	■			■
	Sea level rise and coastal erosion	Produce relative sea level projections under different emissions scenarios for each coastal region.	■						
		Assess at-risk facilities due to sea level rise.	■	■	■	■			■

Effect	Strategy				
Greater coastal storm strength with sea level rise	Harden seaside and shore-based facilities.	■	■	■	
More extreme, or more frequent, coastal flooding	Analyze transportation system vulnerabilities in light of storm surge potential.	■	■	■	
	Revise federal, state, and professional engineering guidelines to reflect current and anticipated future climate changes (e.g., precipitation intensity and duration curves) and require their use as a condition for federal investments in infrastructure and incorporate climate.	■	■	■	
	Require climate change assessments in long-range transportation planning in floodplains, and in land use planning in flood-prone coastal areas.	■	■		
	Include climate change considerations in planning within metropolitan planning organizations.		■		
	Identify and take constructive action to provide and protect emergency evacuation routes.	■	■	■	
	Revise FEMA flood maps.	■			
	Strengthen port facilities to temporarily withstand flooding and surges.	■	■	■	■
	Elevate structures and resources.	■	■	■	■
	Build or raise seawalls, levees, and dikes for protection.	■	■	■	■
	Build surge barriers to protect vulnerable rivers and adjacent infrastructure.	■	■	■	
	Retrofit to strengthen infrastructure (tie down bridge decks and protect piers against scour).	■	■		
	Protect critical components (tunnels and electrical systems).		■	■	
	Abandon, relocate, or move infrastructure and facilities.		■	■	■

NOTE: Most adaptations are local and need to be tailored to local conditions. The suitability of each adaptation option must therefore be evaluated in the context of local conditions. Where possible, the table refers to assessments and syntheses that consider multiple adaptation options and provide references to specific studies.

SOURCE: Reference citations were abbreviated as follows to conserve space: RFF-PI (Neumann and Price, 2009).

TABLE 3.7 Energy: Examples of ideas about specific options for facilitating energy sector adaptation to climate change and identification of entities best poised to implement each option.

Climate Change	Impact	Possible Adaptation Action	Private sector	Federal	Local Government	NGOs	Individuals
Average temperature rise	Increased demand for cooling, reduced demand for heating	Increase regional electric power generation capacity (4-5), after careful consideration of the impacts of resource plans in the United States on overall emissions; plan for and implement enhanced delivery capacity (RFF-PI); take into account changing patterns of demand (summer-winter, north-south) when planning facilities (RFF-PI).	■				
		R&D to make space cooling and building envelopes more efficient and affordable.		■			
		Lead by example: government agencies can weatherize buildings and manage energy use to reduce cooling demands (CADGS).		■	■		
	More frequent and/or longer heat waves	Ensure that energy requirements of especially vulnerable populations are met, especially during heat waves (4-5).			■	■	

Climate impact	Adaptation strategy					
	Improve efficiency of energy use, especially electricity use at home and in commercial buildings (e.g., energy audits) (4-5); develop contingency planning for probable seasonal electricity supply outages.	■				
Increases in ambient temperature reduce efficiencies and generating capacity of power plants	Address vulnerability to heat waves in transmission and delivery systems (4-5).			■		
	Improve efficiency of power generation and delivery.			■		
	Provide government incentives to study the issue of whether decentralized power production reduces risk (RFF-PI).		■	■		
Changes in precipitation or water availability	Develop electric power generation strategies that are less water-consuming, especially for thermal power plant cooling (e.g., dry cooling and increased cycles of concentration for cooling water) (4-5); develop contingency planning for reduced hydropower generation, especially in regions dependent on snowmelt.		■			
	Accelerate development of low-energy desalination technologies (4-5); develop higher cycles of concentration in cooling water systems (RFF-PI).			■	■	
	Diversify energy sources to provide a more robust portfolio of options.			■	■	
	Establish incentives for water conservation in energy systems, including technology development (4-5), and for integrated water and energy conservation planning.				■	■
Changes in intensity, timing, and location of extreme weather events	Harden infrastructures to withstand increased flood, wind, lightning and other storm-related stress (4-5); consider relocation of infrastructures to less vulnerable regions in longer term (see sea level rise) (4-5).		■			

	Adaptation option				
	Increase resilience to energy interruptions and other threats; expand redundancy in electricity transmission capacity and energy storage capacity.	■		■	
Disruption of energy transmission and transportation due to extreme events	Assess regional energy-sector vulnerability and communicate vulnerabilities; advocate responsible contingency planning.		■		
	Prepare for supply interruptions (e.g., backup systems for emergency facilities, schools, etc.).	■		■	■
Sea level rise	Conduct regional analysis of vulnerability of coastal energy infrastructure to sea level rise; advocate responsible land use planning and contingency planning.	■		■	■

NOTE: Most adaptations are local and need to be tailored to local conditions. The suitability of each adaptation option must therefore be evaluated in the context of local conditions. Where possible, the table refers to assessments and syntheses that consider multiple adaptation options and provide references to specific studies.
SOURCE: Reference citations were abbreviated as follows to conserve space: 4-5 (CCSP, 2007), RFF-PI (Neumann and Price, 2009), CADGS (California Energy Commission, 2009).

TABLE 3.8 Oceans and coasts: Examples of ideas about specific options for facilitating ocean and coastal sector adaptation to climate change and identification of entities best poised to implement each option.

Climate Change	Impact	Possible Adaptation Action	Federal	State	Local Government	Private Sector	NGO/Individuals
Accelerated sea level rise and lake level changes	Gradual inundation of low-lying land; loss of coastal habitats, especially coastal wetlands; saltwater intrusion into coastal aquifers and rivers; increased shoreline erosion and loss of barrier islands; changes in navigational conditions	Site and design all future public works projects to take into account projections for sea level rise.	■	■	■		
		Eliminate public subsidies for future development in high hazard areas along the coast.	■	■			
		Develop strong, well-planned, shoreline retreat or relocation plans and programs (public infrastructure and private properties), and poststorm redevelopment plans.		■	■		
		Retrofit and protect public infrastructure (stormwater and wastewater systems, energy facilities, roads, causeways, ports, bridges, etc.).	■	■	■		
		Adapt infrastructure and dredging to cope with altered water levels.	■	■			
		Use natural shorelines, setbacks, and buffer zones to allow inland migration of shore habitats and barrier islands over time (e.g., dunes and forested buffers mitigate storm damage and erosion).	■	■	■		■
		Encourage alternatives to shoreline "armoring" through "living shorelines" (NRC).	■	■	■		

Climate change impact	Potential impact	Develop strategic property acquisition programs to discourage development in hazardous areas, encourage relocation, and/or allow for inland migration of intertidal habitats.	Plan and manage ecosystems to encourage adaptation (see ecosystem options).	Facilitate inland migration and relocation of coastal communities.	Strengthen and implement building codes that make existing buildings more resilient to storm damage along the coast.	Increase building "free board" above base flood elevation	Identify and improve evacuation routes in low-lying areas (e.g., causeways to coastal islands).	Improve storm readiness for harbors and marinas.	Establish marine debris reduction strategy.	Establish and enforce shoreline setback requirements.	Reduce CO_2 emissions (Limiting).	Support ocean observation and long-term monitoring programs.	Evaluate and manage for ecosystem and infrastructure impacts.
Changes in sea ice	Changes in ecosystem structures	■		■		■	■		■		■	■	■
Changes in sea ice	Exacerbate coastal erosion; severe storms reach coast	■	■	■	■	■	■	■	■	■	■		
Increased intensity/frequency coastal storms	Increased storm surge and flooding; increased wind damage; sudden coastal/shoreline alterations		■	■	■	■	■	■	■		■		
Ocean acidification	Potential changes in ocean productivity and food web linkages; degradation of corals, shellfish, and other shelled organisms; potential impacts on coastal infrastructure (i.e., construction materials)			■				■			■		

Category	Potential impacts	Potential adaptation strategies					
Changes in physical and chemical characteristics of marine systems	Changes in salinity; changes in circulation; changes in seawater temperature; changes in salinity and temperature stratification; changes in estuarine structure and processes (e.g., salt wedge migration); changes in ecosystem structure ("invasive," nonnative species), species distributions, population genetics, and life history strategies (including migratory routes for protected and commercially important species); increased frequency and extent of harmful algal blooms and coastal hypoxia events	Establish monitoring and mapping efforts to measure changes in physical, biological, and chemical conditions along the coast.	■			■	■
		Utilize approaches that do not endanger species that are harvested or endangered.	■			■	■
		Ensure flexibility in management plans to account for changes in species distributions and abundance.				■	■
		Implement early warning and notification systems for shellfish and beach closures, salinity intrusion in coastal rivers (for industry impacts and water resource management, i.e. freshwater intakes), and for unusual events such as hypoxia.	■	■		■	■
Changes in precipitation	Increased runoff and non-point source pollution or eutrophication; changes in coastal hydrology and related ecosystem impacts; increased coastal flooding	Improve non-point source pollution prevention programs.				■	■
		Improve stormwater management systems and infrastructure.			■	■	
		Improve early warning systems for beach and shellfish closures.	■	■	■	■	■

NOTE: Most adaptations are local and need to be tailored to local conditions. The suitability of each adaptation option must therefore be evaluated in the context of local conditions. Where possible, the table refers to assessments and syntheses that consider multiple adaptation options and provide references to specific studies.

SOURCE: Reference citations are abbreviated as follows to conserve space: NRC (NRC, 2007c), Limiting (ACC: *Limiting the Magnitude of Future Climate Change* [NRC, 2010c]).

CHAPTER FOUR

Managing the Climate Challenge: A Strategy for Adaptation

As the previous chapter demonstrates, many ideas are available about ways to adapt to climate variability and change. However, few of these options have been assessed for their effectiveness under projected future climate conditions and how they might interact across the sectors and with other stressors. In addition, effective planned adaptation is hampered by a number of challenges and barriers, and there is a need for a comprehensive approach to adaptation planning to overcome them. In the face of both limited knowledge on adapting in the context of climate change and the importance of addressing climate change risks in a prudent and timely manner, a process is needed that allows decision makers to identify and address the most urgent risks and incorporate new information and knowledge in the decision-making process in an iterative fashion. This chapter reviews the challenges and barriers and suggests some approaches to choosing among the many options to manage the risks associated with climate change, using the example of New York City's recent adaptation activities.

THE ADAPTATION CHALLENGE

Despite the nation's substantial economic assets, at present its adaptive capacity to respond to new stresses associated with climate change is limited. As a starting point, it can be argued that our societies are not even well adapted to the *existing* climate, especially to well-understood natural hazards (hurricanes, floods, and drought) that continue to result in human disasters (Mileti and Gailus, 2005; NRC, 2006; O'Brien et al., 2006). Numerous reports and academic research papers describe long-standing impediments to natural hazards mitigation, and these challenges will continue to limit our capacity to adapt to climate change—especially when it involves the intensification of natural hazards (NRC, 2006).

Adaptation requires both actions to address chronic, gradual, long-term changes such as ecosystem shifts and sea level rise, and actions to address natural hazards that may become more intense or frequent. Addressing gradual changes is challenging because the eventual extent of such changes is difficult to recognize and measure, plans beyond 20 years are usually met with skepticism, and costs for initial investments may be unaffordable even when cost-effective in the long term. The experience of New Orleans with Hurricane Katrina—and in fact, continued development throughout the nation in hazardous areas that increase exposure to coastal storms, flooding, and wildfires—indicates a need for fundamental changes in the management of climate-sensitive resources such as coastal areas regardless of the intensification of hazards due to climate change. The continued development of vulnerable areas such as those prone to flooding increases climate risks. Climate changes such as sea level rise and increased storm intensity further exacerbate climate risks, bringing new urgency to these issues. Existing recommendations for improvements in natural hazards management (Heinz Center, 2000; NRC, 2006) should be considered very seriously since many of these actions would address the most immediate needs for climate adaptation as well.

Political Impediments

For several decades, climate change adaptation has been neglected in the United States, perhaps because it was perceived as secondary in importance to mitigation of greenhouse gas (GHG) emissions, or perhaps more importantly because it would actually take attention away from mitigation by implying that the country can simply adapt to future changes (Adger et al., 2009; Moser, 2009b). In addition, the topic of climate change and the discussion of options for responding have become much more highly politicized in the United States than in some other parts of the world. Arguments in the media over whether climate change is "real" and to what degree it is a problem generated by human activity have confused people about whether action is needed and whether their actions can make any difference (Boykoff, 2007). Furthermore, there are frequent suggestions in the media that responding to climate change is "too expensive" or that the options available to limit emissions and adapt to impacts will have a negative impact on the U.S. economy.

Adaptations to long-term problems involve long-term investments and bring considerations of intergenerational equity and other social and economic factors into play that significantly affect the calculation of costs and benefits. The influences of climate change extend well beyond the election cycle of the typical public official in the United States. Long-term adaptations must, therefore, hold some promise of short-term reward if they are to be attractive to elected decision makers.

Institutional and Resource Limitations

Several reports have found that current U.S. institutions at virtually every scale lack the mandate, the information, and/or the professional capacity to select and implement climate change adaptations that will reduce risk sufficiently, even when these adaptation actions are urgently needed (Moser, 2009b; NRC, 2009a,b). New institutions and bridging organizations will be needed to facilitate the communication and integrated planning required to address complex intersectoral problems that cross geographic scales (*ACC: Informing an Effective Response to Climate Change*; NRC, 2010a). Moreover, the availability of funding for climate change adaptation at most levels of government has been highly constrained, and there are few public-sector entities that have identified resources for adaptation (NRC, 2009a).

Identifying new financial resources that can be directed toward adaptation might be difficult in any case, but it is particularly challenging as the world's major economies struggle to recover from the worst recession in decades. The vagaries of economic cycles and the associated political volatility make it clear that adaptation efforts need consistent sources of funding over time because "stop-and-go" efforts are far more expensive and far less effective. Mainstreaming adaptation considerations and outcomes into decisions with climate-sensitive consequences (such as reauthorization of laws affecting land and water use, the National Flood Insurance Program, or the Coastal Zone Management Act) is one way to reduce cost, provide incentives to adaptation, and perhaps smooth the intensity of adaptation efforts (see Box 4.8 later in this chapter).

Notwithstanding efforts to reduce costs, the total expenditures on adaptation will most likely have to be substantial and grow over time. There is very little reliable source material, however, on the total financial costs of adaptation, particularly for the United States. To be sure, some studies apply uniform and, in many cases, simple rules to estimate how societies will adapt and the cost of such adaptations. Such a "top-down" approach often does not sufficiently account for geographic variation in vulnerabilities, adaptations, and costs, and it usually fails to distinguish between voluntary and policy-driven adaptation. Some recent studies have attempted to estimate either global costs of adaptation or total costs for developing countries. For example, the United Nations Framework Convention on Climate Change (UNFCCC, 2007) estimates annual costs of $49 to $170 billion for global adaptation by 2030, and the World Bank (2009) estimates annual costs of $75 to $100 billion by 2050 in developing countries alone. Parry et al. (2009) concluded that the UNFCCC (2007) estimate may be too low by a factor of 2 to 3.

The literature does not contain comprehensive estimates of adaptation costs in the United States, but early estimates for some sectors have been published. For example, it has been estimated that the *cumulative* infrastructure costs of protecting low-lying coastal areas in the United States from up to a 3-foot sea level rise could reach more than $100 billion (Neumann et al., 2000), which avoids even larger losses to property and land values. The cumulative costs of adapting water resource infrastructure to climate change in 2050 are estimated to be half a trillion dollars (CH2MHill, 2009). These studies suggest that the annual financial costs of worthwhile adaptation in the United States could be tens of billions of dollars by midcentury.

Although the lack of funding is a much more serious concern in developing countries, it is clear that the United States has failed to properly maintain existing water, wastewater, transportation, and energy infrastructure even for the climate that it faces now (see AWWA, March 2009,[1] estimates of infrastructure repair needs). As a result, there is already an "adaptation deficit." The need to cope with a dynamic climate that will pose new threats over time only adds to the challenge and will most likely increase the costs of investing in infrastructure.

MANAGING THE RISK

Adaptation is fundamentally a risk-management strategy. Risk is a combination of the magnitude of the potential consequences and the likelihood that they will occur. Managing risk in the context of adapting to climate change involves using the best available social and physical science to understand the likelihood of climate impacts and their associated consequences, then selecting and implementing the response options that seem most effective. Because knowledge about future impacts and the effectiveness of response options will evolve, policy decisions to manage risk can be improved if they incorporate the concept of "adaptive management," which implies monitoring of progress in real time and changing management practices based on learning and as a recognition of changing conditions (*ACC: Informing an Effective Response to Climate Change*, NRC, 2010a; NRC, 2009a; NRC, 2004) is incorporated. The National Research Council (NRC, 2009a, p. 76) report states, "Rather than presuming that managers make one-time decisions on the basis of the best existing knowledge,

---

[1] The U.S. Environmental Protection Agency's quadrennial assessment now estimates that $334.8 billion needs to be spent over the next 20 years on drinking water infrastructure needs. The American Society of Civil Engineers (ASCE) *2009 Report Card for America's Infrastructure*, which gave drinking water and wastewater infrastructure a grade of D−, cited investment needs totaling around $1 trillion for both water and wastewater over the same period. ASCE estimates the funding shortfall on drinking water projects alone will be $11 billion annually.

adaptive management regards policy choices for complex environmental problems as part of a carefully planned, iterative, sequential series that emphasizes monitoring and learning as the system changes, both in response to external stimuli and in response to managers' actions."[2]

This section proposes a framework to manage the risk associated with the impacts of climate change on the natural and built environments. The framework includes (1) identifying the key problems and asking the right questions, (2) assessing the risk, (3) perceiving the risk, (4) properly communicating risk to decision makers, and (5) designing and implementing risk-management strategies.

Identifying the Problems and Asking the Right Questions

It is important to be clear at the start about the problem to be managed. Without a shared understanding of the nature of the problem, the desired goals of the stakeholders, and the "decision context" (Jacobs et al., 2005), collective risk management is not likely to be successful. In framing the problem, it is important to include the perspectives of individuals and interested parties whose voices and concerns might not otherwise be heard, those who will assume the responsibility of administering and implementing the adaptation and sustaining it over time, and those involved in monitoring success or failure against stated goals and objectives. In coping with uncertainty, it will be particularly important to separate relevant signals from random noise in the observations, carefully analyze new scientific information, and design "midcourse" adjustments based on lessons learned.

For example, when developing adaptation measures to reduce the impact of sea level rise on damage to coastal areas from floods and hurricanes, key interested parties should include the relevant public- and private-sector agencies concerned with climate change impacts, businesses that will develop technology or approaches to adaptation, those who are vulnerable economically and physically (e.g., adverse health or environmental effects), and those who will have to pay for adaptation measures and deal with the adverse impacts from global warming.

---

[2] It is important to note that "adaptive management" is used here in its most general form. It implies an iterative process in which decisions are based on evolving understanding of the underlying natural and social science and the observed success (or failure) of programs and policies that have been implemented. The panel is not using the Holling (1978) framework, in which policies and programs are viewed as experiments designed to elicit new information.

Assessing Risk

In the context of climate adaptation, assessing risk means identifying specific (climate-related) events and evaluating their potential adverse (and in some cases beneficial) consequences in terms of magnitude, spatial scale, time frame, duration, intensity, and consequences for society. Risk as assessed by experts encompasses studies that estimate the chances of a specific set of events occurring and/or their potential consequences (Haimes, 1998). The primary goal of assessing risk is to produce information that improves risk-management decisions and to identify and quantify the impact of alternative actions (including the status quo) and their consequences. Assessing risk may also include considering the nature of vulnerabilities and consequences associated with specific risk-management decisions.

Once the problem is well identified, assessing risk begins with hazard identification— the process of specifying the scope of the assessment. In the case of climate change, the available empirical evidence can be summarized with respect to its potential impacts on natural and social systems and different economic sectors, including interactions between sectors and systems. In some cases, these impacts can be associated with specific climate futures (high, medium, and low emissions over time, with or without efforts to limit atmospheric GHG concentrations). Other scenarios can be based on emissions trajectories with distributions of impacts over time. In either case, the process of assessing risks will vary depending on the sector being examined and the interests and concerns of affected stakeholders.

Scenario analysis is a widely used technique for identifying vulnerabilities from climate change (see Mearns and Hulme, 2001). In some cases, the relative likelihoods of alternative futures can be supported using risk-based techniques such as expected cost-benefit analysis and decision analysis, including expected value of information and efficient risk-spreading designs (e.g., insurance). In other cases, several scenarios (e.g., high, median, and low) can be employed to span the range of possible outcomes. Here, the robustness of alternative responses and the potential value of hedging strategies can be explored, but only if the scenarios capture a wide range of changes in key climate variables such as temperature, precipitation, and sea level rise. The diversity of possible futures is particularly critical to explore if the direction of change in a key variable such as precipitation is not clear (for example, in cases where climate models do not give consistent projections or predict whether precipitation increases or decreases).

If adaptations are expensive, decision makers might want to judge the appropriate timing for incurring the costs of adaptation investments. In the context of coastal flooding, for example, property owners might want to explore the frequency of other sources of similar vulnerability (e.g., extreme local precipitation events) to judge potential ancillary benefits of adaptation investments. For example, elevating houses in the face of rising water will not only reduce the potential losses to one's own property but also alleviate damage to neighboring structures. Effects of climate change, including rises in sea level, increases in storm frequency, and changes in the ecological properties of natural systems (such as loss of storm-buffering wetlands or mangroves), not only will impact coastal infrastructure but will also alter potential tradeoffs between different approaches to reducing risk (CCSP, 2009b).

Perceiving Risk

Risk perception is concerned with the psychological and emotional factors (e.g., anxiety, regret, and peace of mind) that have been shown to have an enormous impact on behavior and that need to be considered when developing risk-management strategies (Slovic, 2000). Decisions are affected by the perceptions of those who make them, so the potential importance of risk perception cannot be overestimated. There is a wide disparity between the views held by the general citizenry and those of experts about risks associated with climate change—both the nature and seriousness of the consequences and their associated probabilities (Leiserowitz, 2005). There is also a growing body of evidence showing that emotions play an important role in individual's decision-making processes. Rather than basing one's choices simply on the probability and the consequences of different events, as normative models of decision making would suggest, individuals are also influenced by emotional factors such as fear, worry, and love (Finucane et al., 2000; Loewenstein et al., 2001). These concerns deserve serious consideration when developing adaptation strategies for addressing the future impacts of climate change. In addition, an important factor in motivating protective actions by individuals is their knowledge about what to do to reduce one's vulnerability to a certain hazard (Paton, 2008). These factors should be anticipated in discussions about the appropriate decision-support approach and taken into account when developing and communicating risk and managing risk strategies. Educational programs may, in these cases, be prerequisites for galvanizing public support for expensive but important, forward-looking responses (*ACC: Informing an Effective Response to Climate Change*; NRC, 2010a).

Communicating Risk

Well-designed approaches to explaining risk encourage the effective participation and interaction of technical experts, stakeholders, and decision makers in managing risk decision processes and deliberations. Poorly designed communication can breed resentment and mistrust (NRC, 1996b), especially when communication techniques are perceived as biased or inappropriate to the audience. A growing number of studies have shown that communication of risk and uncertainty is important in helping people respond to climate change (NRC, 1996b, 2010a). What most risk researchers consider the ideal approach for communicating uncertainty and risk focuses on establishing an iterative dialogue between stakeholders and experts, where the experts can explain uncertainty and the ways it is likely to be misinterpreted; the stakeholders in turn can explain their decision-making criteria as well as their own local knowledge in the area of concern; and the various parties can work together to design a risk-management strategy, answering each others' questions and concerns in an iterative fashion (Fischhoff, 1996; NRC, 1996b; Patt and Dessai, 2005).

Designing and Implementing Risk-Management Strategies

Based on case studies and its own experience, the panel suggests that providing a portfolio of options for managing risk is likely to be more effective than relying on a single solution. Using multiple strategies simultaneously—such as providing public education (i.e., communicating risk), offering economic incentives (e.g., subsidies, fines, or tax incentives), or intervening directly to prevent or avoid consequences (e.g., by implementing regulatory standards or restricting activity)—provides the most robust way to address risks. The multiple strategies chosen sometimes involve public-sector actions (e.g., regulations, standards); in other circumstances, they can include strategies to spread or transfer risk (e.g., offering or requiring insurance and/or compensating for losses). Strategies for managing the risk can be evaluated in a variety of ways (see detailed discussion below). "Robust decision making" tests a number of different options and results in a decision path that keeps as many future options on the table as possible (*ACC: Informing an Effective Response to Climate Change*; NRC, 2010a).

Currently, however, there is inadequate information about the effectiveness of adaptation options in many specific applications, frequently because essential metrics for evaluation are unavailable or because the responsible party perceives the benefits from monitoring outcomes to be insufficient to justify this activity. For now, much can be learned through discussions, collaboration, and applying lessons learned from experiences at the local or state level or in other countries. Learning by doing, as well as by deliberately testing different adaptation approaches (e.g., NRC, 2009a), sustained monitoring, collection, and analysis of the right data will be essential if we are to identify lessons and evaluate effectiveness. Furthermore, it is challenging to assess the net value of an adaptation measure given the uncertainties associated with changes in climate and impacts as well as the long time frames over which climate change will happen. Designing assessment programs to evaluate adaptation costs and successes critically is therefore essential. Research in this area should recognize that adaptations will likely need to be adjusted as the climate continues to change and that, in some cases, the actual benefits of adaptations may not be obvious or measurable for many decades following the investment.

There is also a need for risk-management approaches that accommodate multiple metrics (standards of measurement) of climate impacts, and for decision-support tools that complement more traditional cost-benefit approaches of policy analysis (see, e.g., Yohe, 2009a,b). Many impacts (on ecosystems, for example) and many contexts (including social consequences) cannot be fully monetized (Downing and Watkiss, 2003). If decisions are made comparing only economic measures quantified exclusively in financial terms, then social and ecological consequences and other nonmonetized impacts cannot be considered in proportion to their significance.

While it is essential to communicate the risks of inaction, it is equally essential that the risks of any potential adaptation action—including indirect consequences—be effectively communicated. For example, those residing in areas that are protected by dams or levees are likely to believe that they are fully protected against future flooding or storm surge. In reality, some levees have been poorly designed, as evidenced by Hurricane Katrina, but it is only after a hurricane or flood that attention is drawn to these inadequacies. Furthermore, there is likely to be new development in these regions unless officials and the public are made aware that it is possible for these flood-control and engineering solutions to fail. The result, known as the "levee effect," can be losses that are much larger than they would have been if the risks had been correctly perceived and economic development in these areas had been limited in the first place (Tobin and Montz, 1997).

As noted above, perceptions of risk are often inconsistent with scientific approaches to assessing risk. Factors such as fear and anxiety impact judgments of risk, and shortsighted behavior is common, making it particularly important to design adaptation strategies that encourage individuals to take a longer-term perspective in their own best interest (and that of society more generally). Given the tendency of decision makers to evaluate the net benefits of strategies over only a 2- to 3-year span, it may be necessary to provide short-term returns to encourage the adoption of long-term adaptation strategies. For example, long-term contracts with short-term returns may be required to encourage investment in cost-effective adaptation measures for dealing with increasing damage from future floods or hurricanes as sea level rises (see Box 4.8, later in this chapter). In this case, long-term flood insurance (to spread risk) and long-term home improvement loans (to finance adaptation measures) might be combined with initial discounts on annual premiums that are likely to exceed the annual cost of the loan (Kunreuther and Michel-Kerjan, 2009).

A complementary approach is to revise and enforce land use regulations and building codes using standards that reflect the risks of climate change impacts. In the case of building codes, the challenge is to apply them to existing structures in harm's way rather than just new structures. Often such retroactive codes are necessary, particularly when property owners are not inclined to invest in adaptation measures on their own—either because they perceive a disaster will not happen to them or they believe the risk is small relative to the cost of adaptation. Insurance premium reductions coupled with long-term loans could make well-enforced building codes more palatable to property owners.

DEVELOPING AN EFFECTIVE ADAPTATION STRATEGY

Because the perception that climate fluctuates around a stationary mean is in conflict with recently observed climate dynamics (Milly et al., 2008), decision makers need an approach that is responsive to changes in the likelihood of extreme outcomes as well as changes in the "average" climate. Resources and natural processes may change their function and their location, and in some cases may cease to exist (IPCC, 2007a; West et al., 2009). Rather than managing the resource to maintain its past condition and state, management may need to take steps to protect the resource (e.g., building coastal defenses) or allow the resource to change as needed to adapt to climate change (e.g., allow migration of species to new habitats or manage forests for fire control instead of for timber). In other words, the managers of these resources must work to incorporate the impact of climate change in their plans and operations. Instead of managing for resilience (which implies returning to the status quo), managing for change might become a more feasible approach (West et al., 2009).

This chapter uses the recent example of New York City (see Box 4.1), one of the most comprehensive approaches so far to adaptation in the United States, to illustrate some basic principles for developing and implementing an effective adaptation strategy. There are a number of other examples of municipalities and states (large and small) that have developed and begun implementing adaptation strategies. Keene, New

---

BOX 4.1
The New York Experience

New York City responded to the challenges of a changing climate by creating several participatory mechanisms: broad public input through PlaNYC, a Climate Change Adaptation Task Force, and a Policy Working Group. PlaNYC was announced in December 2006 as a sustainability and growth management initiative to answer the question of what kind of city New York should be—posing the question "Will you still love New York in 2030?" After PlaNYC was announced, a top-to-bottom outreach effort began to receive input on goals for 2030. Staff met with thousands of community leaders and representatives and received thousands of e-mail suggestions. This input was synthesized into "10 goals for 2030," which became the basis for PlaNYC.

The plan's scope was soon expanded to consider climate change as a major threat to sustainability. The 10 goals of PlaNYC are distributed across three challenges areas (growth, infrastructure, and the environment) and six planning elements that correspond to the city's environment: land, water, transportation, energy, air, and climate change.[1]

Responding to strong leadership from the mayor's office under Michael Bloomberg, New York City created a Climate Change Adaptation Task Force that included various private and public stakeholders. The stakeholders were divided into working groups that represent broad categories of infrastructure: communications, energy, transportation, water, and waste. In addition, a Policy Working Group was convened to review the codes, rules, and regulations that govern infrastructure in New York City and to identify those that may need to be changed or created to help the city cope with climate change. Each working group provided a forum within which stakeholders could identify common vulnerabilities, share best practices, take advantage of potential synergies, and develop coordinated adaptation plans. At appropriate stages in the Task Force process, the individual working groups came together to ensure consistency, identify opportunities for coordination, investigate impacts of adaptation strategies on other sectors, and develop cross-sector adaptation strategies. Since the Policy Working Group's focus on the regulatory context spanned a broad range of issues, it was particularly vigilant in coordinating its efforts with other working groups. Adaptation strategies that impacted multiple stakeholders were identified as possible citywide strategies and forwarded to the full Task Force for further evaluation and development.

---

[1]For details, consult *http://www.nyc.gov/html/planyc2030/html/plan/plan.shtml*.

Hampshire (ICLEI, 2007), is, for example, a small municipality that is carrying out a similar effort in a much different context. The panel focuses initially on New York, however, because the city has documented its steps very carefully, including peer review, in designing a plan that incorporates the impacts of climate change as a threat to sustainability, and it has worked to create decision-support templates that may be useful for others.

New York City's Climate Change Adaptation Task Force used a common set of climate change projections that were developed with support from the Rockefeller Foundation by the New York Panel on Climate Change (NPCC) in the Climate Risk Information workbook (NPCC, 2009). The NPCC projections allowed stakeholders to gain an understanding from the outset of the climate science, initial potential impacts derived from existing and evolving climate variability, and uncertainties. This ensured that the inventories of infrastructure at risk and associated adaptation strategies were based on the same state-of-the-art climate change projections. Two key aspects of an adaptation strategy—organization and objectives—are explored in the following sections.

Organization of Adaptation Efforts

Snover (2007) outlines steps that governments can follow to adapt to climate change, including establishing public engagement and planning processes. Developing an adaptation strategy can require involvement of many stakeholders (Smith et al., 2009b), as the New York City experience has shown. In government, all departments managing affected resources need to be involved in developing the strategy, including budget and legal departments. Stakeholders who will be affected by climate change or who will be involved in implementing adaptations should also have a role, and each group is likely to bring different values, objectives, and perspectives to the process.

Since stakeholders are likely to have diverse objectives, clear leadership is needed from the outset in the development of adaptation strategies. Direct input and engagement of a chief executive of the government is typically required to make a clear and pressing case that adaptation is important. The development and implementation of the strategy can be directed from the chief executive's office or coordinated by a lead agency. The advantage of managing it from the chief executive's office is that it can demonstrate the chief executive's commitment and avoid turf battles among departments. On the other hand, running a coordinated effort out of a larger department might provide access to more staff and resources typically affiliated with a department compared to the chief executive's office.

Defining Objectives

Defining objectives is a key part of developing any climate change adaptation strategy (Burton, 2008; King County, 2007; Snover, 2007; Theoharides et al., 2009). Objectives can be tied to goals and used to define metrics for measuring success and monitoring progress. Clear goals facilitate setting criteria for monitoring the success of programs that are designed to meet new challenges over time (NRC, 2005b). If goals and priorities are broadly stated and not clearly defined, developing strategies to meet them will be more difficult. For example, goals such as "protect human health and welfare," while laudable, may not give sufficient precision to develop specific metrics. Such a goal may conflict with another goal, "develop levels of protection that are not too costly." Defining acceptable levels of risk for planning purposes—for example, risk of mortality of one in a million or risk of property damage from an extreme event of one in a hundred—can be justified in the context of precise and prespecified goals (NRC, 2005b).

Decision makers often construct or select plans that are designed to achieve multiple goals. For example, a plan to invest in loss-reduction measures (e.g., elevating a structure) to reduce the magnitude of damage from a future flood may have three goals: (1) reducing the magnitude of a catastrophic loss, (2) reducing anxiety about the possibility of a severe loss, and (3) not incurring a large up-front cost for instituting the measure. The decision on whether to undertake an adaptive measure depends on the tradeoffs between values assigned to these sometimes-conflicting goals and the likelihood of achieving each goal (Krantz and Kunreuther, 2007).

Ideally, the adaptation planning process begins by defining goals and objectives, then examining options to meet these goals in terms of their impacts on different stakeholders. A fundamental issue to address is whether preexisting strategic goals will continue to be appropriate or feasible in light of anticipated climate change impacts or whether these former priorities will need to be modified. This issue centers around whether adaptation will allow the goals to be met as before or whether goals need to be adjusted because climate change alters the feasibility of achieving them. For example, a preexisting goal of maintaining agricultural productivity in a region may still be achievable under climate change if farmers can modify management practices or use more inputs such as herbicides or water. In those cases, farming might still be profitable, but with reduced profit. In other circumstances, however, farmers in the region may need to change crops to meet changing climate conditions, because climate change has reduced the productivity of old crops. In still other cases, farming in the region may not remain feasible in the long term, and the land might need to be retired from farming. Whether a preexisting goal can be maintained or needs to be modified will depend in each case on specific local circumstances. The relative importance of a previous goal may also change when climate change is considered.

The discussion above suggests that an overriding goal may be one of accommodating

BOX 4.2
New York City's Objectives

New York City chose to define its objectives in terms of the "acceptable levels of risk" that are embodied in its current codes and regulations. While it is difficult to quantify society's acceptable level of risk, it turned out to be relatively easy to identify certain things that New Yorkers would perceive as unacceptable, such as allowing the New York City subway system to flood multiple times per month. The city also recognized that society's acceptable levels of risk could change over time as the climate changes. Indeed, if adjustments in acceptable risk levels were not made for climate changes, then associated changes in climate variability would likely result in more frequent episodes of intolerable consequence. It became clear, as a result, that accepting risk as a metrical implied that an adaptation process should create "flexible adaptation pathways" that would allow policy makers, stakeholders, and experts to develop and implement strategies that evolve as climate change progresses. Moreover, it became clear that flexible adaptation pathways can only be constructed on the basis of solid understandings of current and future climate hazards, rigorous evaluations of regulations and design standards, and interactive decision-support tools that help infrastructure managers to plan adaptation strategies. Monitoring and reassessment of climate science, adaptation strategies, and policies are critical as well, so that responses to the evolving risks of climate change can be adjusted effectively (NPCC, 2009).

change (see Box 4.2), either directly, by implementing adaptation options, or indirectly, by changing management goals and expectations if climate change makes them no longer achievable (West et al., 2009).

DEVELOPING AN ADAPTATION PLAN

Like other planning processes, adaptation planning includes identifying potential problems, developing options to address them, ranking them according to criteria, and selecting the ones to be implemented. Climate change adaptation should, if possible, be incorporated into established planning activities, or "mainstreamed," as New York City has done, rather than requiring a separate approach. There may be common solutions to some problems, but no one approach will be appropriate for all applications. Whereas Chapter 3 discusses adaptation options in depth, this section briefly discusses approaches for identifying and selecting options.

Given the realities of ongoing changes in climate and uncertainty about the effectiveness of currently available adaptation options, the most important principles behind effective adaptation planning are flexibility and an adaptive approach designed to

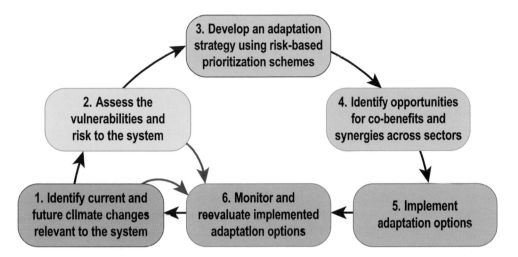

FIGURE 4.1 The planning process is envisioned to incorporate the following steps: (1) identify current and future climate changes relevant to the system, (2) assess the vulnerabilities and risk to the system, (3) develop an adaptation strategy using risk-based prioritization schemes, (4) identify opportunities for co-benefits and synergies across sectors, (5) implement adaptation options, and (6) monitor and reevaluate implemented adaptation options.

meet objectives and goals under a wide variety of future climate conditions (Glantz, 1988; Smith, 1997). To build flexibility and an adaptive approach into the planning process, monitoring and reevaluating the implementation is a key element (Figure 4.1). Since future emissions of GHGs, climate sensitivity (i.e., how much average global temperatures will increase with a given concentration of GHG), and regional patterns of changes in climate are not certain, we cannot develop adaptations now for a specific future climate. Groves and Lempert (2007) used the term "deep uncertainty" to describe the wide range of climate conditions planners and others must cope with. Therefore, monitoring of the climate changes, vulnerabilities, and impacts is also necessary and subsequently requires decision makers to reevaluate and update the planning process (Figure 4.1). The adaptations need to meet goals and objectives across this wide range of climate conditions.

Identifying Adaptation Options

Hence the principle of flexibility: adaptations must perform well under a variety of future climate conditions. This means that adaptive responses must either be able to change to keep pace with different climate conditions or be able to absorb a wide range of climate conditions. Changing in response to changing climate conditions may involve being responsive to climate and other signals. For example, market mechanisms such as prices can be responsive to changing conditions. Using water markets to set water rates can allow for rates to rise when water supplies are tight (e.g., during droughts) and fall when supplies are abundant (e.g., during wet periods). Such adaptation mechanisms can be described as being resilient. A key issue for climate change adaptation is the capacity of the affected system to recognize and quickly respond to signals of changes in climate. (Note the approach need not be cognizant of changes in climate, just changes in conditions. The price of water might rise as supplies tighten, regardless of peoples' perceptions about climate or the causes of a current drought.)

The second case—designing adaptations that can absorb a wide range of climate conditions—means the affected system must be able to continue functioning as intended even in extremes. This case may best apply to design of infrastructure such as water supply, treatment, flood protection, and roads. Such infrastructures are typically designed to be robust—to withstand extreme events and continue to provide intended services.

Climate adaptations can fall into either of two categories: "no regrets" or "climate justified." No-regrets strategies are ones that can be justified without taking into account climate change but which also meet goals and objectives of adaptation to climate change. Examples of no-regrets adaptations include reforming insurance regulations to enable insurers to set rates that reflect risks. (Despite the label, "no-regrets" policies can entail winners and losers: they do not mean that all stakeholders are better off or satisfied.) Climate-justified adaptations, on the other hand, are those that are undertaken mainly because climate is changing or anticipated to change. For example, this might involve raising a flood barrier in anticipation of future increases in sea level or storm intensity (Reeder et al., 2009).

Each of these approaches may be both complex and difficult to implement. For example, no-regrets strategies may perform well with modest changes in climate but may be inadequate for extreme changes. In the case of market mechanisms, the price of water could become unaffordable for low-income households in extreme droughts if set by the market; also, market mechanisms rarely protect environmental interests. Hence, the public sector might have to intervene in some way to ensure equity and provide assistance to those groups harmed by such adaptations.

Box 4.3 describes the process New York City used to identify adaptation options.

BOX 4.3
Identifying Adaptation Options in New York

New York City employed three tools as it worked to identify and analyze adaptation options through the Climate Change Adaptation Task Force:

1. *Infrastructure questionnaires.* Sector-specific questionnaires were designed to help stakeholders create inventories of their infrastructure at risk to climate change impacts, especially those impacts driven by dynamic climate variability.
2. *Risk matrix.* A tool was designed to help stakeholders categorize their lists of at-risk infrastructure based on likelihood of impact and magnitude of consequence.
3. *Strategy framework.* A framing tool was designed to assist stakeholders in developing and prioritizing adaptation strategies based on criteria related to factors such as effectiveness, cost, timing, feasibility, and co-benefits.

Together, these process-based tools provided the foundation for the development of climate change adaptation plans for critical infrastructure in the New York City region. The location of these adaptation plans within an overall planning process is described in detail in the Adaptation Assessment Guidebook created by the NPCC (2010). Perhaps more importantly and similar to Figure 4.1, the process includes monitoring and assessment of results, which feeds directly into subsequent iterations of the same process. It is therefore envisioned as a dynamic cycle of analysis and action followed by reanalysis and possible adjustments to or continuation of previous actions (learn, then act, then learn some more). The New York approach therefore embraces completely the need for flexible adaptation pathways that evolve over time as understanding of the climate as well as the local, national, and global economies change. Indeed, as of the summer of 2009, the Bloomberg administration intended to push the City Council to pass a law that would require each subsequent administration to submit progress reports and revised climate adaptation plans, just as each is now required to submit updates on progress toward sustainability goals.

Comparing and Selecting Adaptation Options

As the previous chapters indicate, many potential adaptation options can be identified. There may be multiple options for the same decision or more options than can be afforded. Thus, options must be prioritized and selected before implementation can begin. This section suggests some criteria that can be used to set priorities and selection criteria for adaptation options and describes methods that can be used to aid decision making.

Considering Benefits of Adaptation

An important factor in setting priorities will be the relative effectiveness of adaptations in reducing adverse impacts (or taking advantage of positive impacts) of climate change. Managed systems such as water resources or agriculture may be more amenable to successful intervention than systems that are subject to relatively little management such as natural ecosystems.

While it may be feasible to compare the effectiveness of various adaptation options within a sector using accepted metrics for that sector, comparing adaptations across different sectors creates methodological issues. To set priorities across sectors, the choice of metrics becomes more complicated because different metrics may be appropriate for different sectors.

Co-benefits

An additional consideration is whether adaptations have co-benefits. That is, does an adaptation create benefits beyond the immediate goals and objectives of adapting a specific system to climate change? For example, protecting coastal wetlands is an adaptation option that may reduce vulnerability to coastal storms. Such wetlands can also provide breeding grounds for fish as well as recreation and tourism amenities, expanding the total benefits of a project. Co-benefits such as these add to the direct benefits of adaptation, raising the relative value of an adaptation option. In fact, many adaptation options listed in the previous chapter currently offer more benefits from adaption to other stressors than to climate change, although it is anticipated that such options also increase resilience to changing climate conditions.

Adverse Impacts

Often overlooked in discussions of adaptation are the potential adverse consequences of adaptation strategies themselves, particularly when they involve engineering solutions. For example, the use of coastal barriers to protect structures or property from erosion as sea level rises blocks natural migration of wetlands and results in loss of beaches (CCSP, 2009b). Similarly, building reservoirs to capture stream flow can alter natural flow patterns and block migration of fish and other aquatic species. Reservoirs also reduce sediment delivery to delta systems, exacerbating beach erosion and impacts from sea level rise (CCSP, 2008a). Often, adaptations in one sector can create adverse impacts in another system or sector. Impacts across multiple sectors can complicate the task of quantifying the adverse impacts in a single or common metric that can be compared with benefits.

Adaptation Costs

Cost is also a factor in setting adaptation priorities. Everything else being equal, the greater the adaptation costs, the less attractive an adaptation option will be. All the costs of implementing adaptations, including capital costs, maintenance and operations over time, reconstruction, and staffing, should be counted. Limitations on availability of staff or availability of staff with the necessary training, experience, or other credentials to implement adaptations may also add to the cost.

Timing of Impacts

The timing of climate change impacts can be a factor used to set priorities in planning for and implementing adaptation (although the uncertainties about the rate and magnitude of climate change discussed in Chapter 2 will also affect projections about the timing of impacts). The following issues influence considerations about timing:

- When will climate change impacts become critical?
- Have impacts already become critical?
- Does the observed impact appear to be clearly linked to climate change or could it be the result of natural variability?
- How much time will be needed to react to, prepare for, or develop adaptation infrastructure to cope with the anticipated impacts?

The question of needed reaction or preparation time is particularly important. If it is possible to quickly react to projected impacts as they happen, then having longer-term adaptation strategies may not be critical. But, if impacts need to be anticipated, then projections of future conditions will be needed, particularly for decisions about large infrastructure projects that can require decades to plan, finance, and build. Long lead times will likely mean preparing for more uncertainty and a wider range of future conditions (see Box 4.4).

As climate continues to change, more and more impacts will become evident, and detection and attribution of impacts will become more important (e.g., DOI, 2008a). Detection is the first step in addressing impacts as they happen, and this will require development of monitoring systems that have sufficient regional coverage and fre-

---

BOX 4.4
Uncertainty and Adaptation Choices: An Example from the Gulf Coast

A recent report by the U.S. Climate Change Science Program, Synthesis and Assessment Product 4.7, *Impacts of Climate Change and Variability on Transportation Systems and Infrastructure: Gulf Coast Study* (CCSP, 2008a), considered vulnerabilities of this region to a combination of sea level rise, more intense storm activity, and land subsidence. According to this report, the region is likely to see apparent sea level rise (actual sea level rise plus land subsidence) of 2 to 4 feet by 2050. Combined with prospects for more severe storms over the decades, this would put significant settlements, infrastructures, and environments at serious risk.

Adaptation options for vulnerable coastal areas are well known (Chapter 3). The main categories are protecting coastal systems with barriers, such as the Netherlands' dikes; hardening coastal systems so that they can handle higher water levels and storm impacts; sharing risks to particular places from low-probability/high-consequence events through insurance; and changing land uses in the region to move vulnerable activities and systems away from vulnerable areas. Some decisions about adaptation approaches seem relatively clear. For example, existing coastal energy and chemical facilities are likely to be diked in order to protect them against another possible Hurricane Katrina in coming years; and coastal cities are being strongly encouraged to improve their emergency preparedness systems.

But deciding what to do in the next several years is shrouded in uncertainties. At what rate will the threats develop? Will the apparent sea level rise be 2 feet or 4? How strong does "more intense" mean? What would be the costs of the various options, and who would bear them? What is likely to be the response of the federal government to a regional challenge that is of national importance?

---

quency of measurement. Just as important is the ability to analyze the information to detect impacts that may be the result of climate change.

Methods for Comparing and Ranking Adaptations

Formal methods available to compare and rank adaptation options include cost-effectiveness analysis, benefit-cost analysis, risk analysis, and multicriteria analysis. These methods differ in terms of information they require and information they provide to decision makers. This section briefly reviews these methods, while acknowledging that none of them are especially good at incorporating social or environmental costs or benefits.

What will be the perspectives of private insurers? What should parties considering new infrastructure investments do?

This is a classic case of a very high-probability threat, whose magnitude and time frame are projected with a high level of confidence, that will challenge America's ability to make decisions under uncertainty. It is deeply imbedded in decisions by the private sector—from energy firms to real estate firms and the insurance industry—and state and local governments. It involves politically charged near-term decisions to protect current settlements, infrastructures, and environments that may not be the right decisions for the longer term.

Addressing these uncertainties in such a complex context involves weighing urgency of action (which in the near term is focused on protection from more intense coastal storms) against resources needed for adaptation. In the near term, coastal infrastructures can be protected by barriers and/or by hardening (e.g., raising roadbeds). In the longer term, however, risk-averse investment strategies are likely to encourage some movement away from some especially vulnerable areas, while hard decisions are made about iconic systems and structures to protect, even at a high cost. In preparation for longer-term participatory strategy discussions, information systems need to provide a steady stream of information about the rate of change in climate change and land subsidence parameters, the rate of voluntary change in land uses in vulnerable areas, and emerging evidence about impacts. And mechanisms need to be developed and used for continuing reevaluation of risks as some uncertainties are reduced.

Making adaptation decisions under uncertainty is in fact an evolutionary process, beginning with relatively low-cost actions that make sense under a range of future conditions, informed by recent experience. It continues with effective information systems to inform further decisions, and it emerges adaptively with a broad-based participatory process reconsidering risk-management strategies as some uncertainties are reduced.

Cost-Effectiveness Analysis

Cost-effectiveness analysis is used to compare alternatives that are expected to achieve the same or a similar goal or benefit. Alternatives are compared based on their relative costs, that is, which alternative costs the least to achieve (approximately) the same goal or outcome. This approach is often applied to examine options for reducing GHG emissions to achieve a certain level of atmospheric GHG concentration. Indeed, this was the point of the "when efficiency" of the Wigley et al. (1996) emissions scenarios—to show that the same concentration limits could be achieved while minimizing the discounted cost of the mitigation interventions over time. Using cost-effectiveness analysis to evaluate adaptation alternatives is appropriate if the objectives or benefits of adaptations are clear and consistent. In many cases, however, there can be multiple benefits of different adaptation measures, making it difficult to make comparisons across alternatives. Alternatives may also have different benefits and can involve a wide range of diverse adverse impacts on other systems (e.g., hard coastal defense structures resulting in loss of beaches or blocking inland migration of wetlands). In such cases, use of cost-effectiveness as a metric for evaluating adaptation alternatives may be inappropriate.

Benefit-Cost Analysis

Benefit-cost analysis (BCA) is a systematic procedure for evaluating alternatives that have an impact on society. BCA attempts to determine which alternatives have the greatest net benefits (difference between benefits and costs) or have higher benefit/cost (B/C) ratio (the ratio of benefits to costs; e.g., Boardman et al., 2001; Loomis, 1993). Each of the alternatives will affect a number of individuals, groups, and organizations in our society. For each alternative, one needs to specify the resulting benefits and costs that impact each of the interested parties. If there is uncertainty associated with the analysis, then one has to assign probabilities to the different states of nature (e.g., floods or hurricanes of different magnitudes) and the resulting outcomes of the interested parties. If the alternative involves multiple time periods, then one has to specify the outcomes that occur in each of these future periods and use a social discount rate to convert these benefits and costs to present value. Once these benefits and costs have been specified, then one needs to quantify these impacts and attach some dollar value to them for each of the affected individuals. BCA uses a common metric—typically money—to compare all benefits and costs in order to determine if benefits exceed costs. The use of monetary metrics has led to criticism that this method fails to capture nonmonetary values such as protection of ecosystems or human life (Brauman et al., 2007). A further disadvantage of the use of money as the common metric is that gains or losses to those societies or individuals with the most wealth can effectively count more than impacts to those with less wealth. This can be addressed by weighting losses to the poor more than the rich (Azar, 1999).

Incongruous timing of when benefits and costs are realized also presents challenges for application of BCA. When costs and benefits are distributed into the future, discounting expresses them all in terms of current dollars. Benefits far into the future or subject to high discount rates will be relatively small in present value terms. BCA can be particularly limited in analyzing benefits and costs over many generations, as may be the case when analyzing the impact of climate change adaptations.

Adaptation measures for reducing the consequences of climate change normally involve an up-front investment cost that provides benefits over a number of years. The nature of these benefits will be a function of the impact that climate change is likely to have on the environment. For example, if global warming produces increasing sea level rise during the next 50 years, then adaptation measures making structures less vulnerable to damage from flooding will have higher discounted expected benefits over the life of the structure than if there was no increase in sea level rise due to global warming.

Due to the uncertainty associated with the nature of climate change over the next 30 to 100 years, it is important to undertake sensitivity analyses to see how robust specific adaptation measures are to different scenarios regarding the nature of climate change. If the net expected benefits or the B/C ratios are positive over a wide range of plausible scenarios, then these adaptation measures would be viewed as desirable.

Risk Management

A risk-management approach to confronting climate change (Carter et al., 2007; Schneider et al., 2007) is gaining traction as a *complementary* analytic tool because it is designed explicitly to make up for many (but by no means all) of the thorny issues associated with BCA. Risk-based analyses rely heavily on information about the relative likelihoods of possible events, which can be challenging to determine in the case of climate change impacts for reasons discussed earlier. Others, including the ones that relate to identifying robust strategies, can be built directly from catalogues of possible futures, even if they cannot be characterized in terms of their relative likelihoods (Yohe, 2009b). Here too, the selection of a metric or metrics can enable or limit comparison of risks across sectors and impacts.

Although there are techniques for doing so, the panel has already noted that assigning probabilities to different climate change outcomes can be challenging. One approach used by some decision makers is not to assign probabilities to different scenarios. In such cases, decision makers may treat different scenarios as equally likely or may focus on the "worst-case" scenarios and ensure that they are prepared for such outcomes.

Risk analysis also incorporates risk perception of individuals, which is normally not considered in undertaking BCA. If decision makers underestimate the value of specific adaptation measures by focusing on their benefits only over the next several years rather than the relevant time horizon (e.g., the projected life of the property where an adaptation measure is considered), then they will underestimate the net benefits of the measure of the B/C ratio. Similarly, if the likelihood of specific events is underestimated relative to the scientific data, then the adaptation measure will be perceived to be less effective than it otherwise would be. There is thus a need to develop risk-management strategies and incentives to encourage the investment in adaptation measures that are viewed as attractive from a benefit-cost approach. (See Box 4.8 for a discussion of innovative risk-management approaches for the National Flood Insurance Program.)

Box 4.5 is devoted to a discussion of the challenges involved in characterizing uncertainty with sufficient clarity that risk-analysis and risk-profiling techniques can be applied in ways that enable one to evaluate the relative attractiveness of different adaptation measures.

Combining consequence and likelihood can provide estimates of the expected values of outcomes. It is often more informative to determine the relative importance of consequences or likelihood in producing high-risk vulnerabilities that should receive the most attention. Box 4.6 describes how New York City considers risk from a very practical perspective that recognizes both its formal definition and the constraints imposed on its application by limited information.

Decision Analysis (Statistical Decision Theory)

Decision analysis (DA) has a structure similar to BCA with two major differences. It normally focuses on a decision made by a specific interested party (e.g., a homeowner, an industrial firm, or a division in an organization) rather than viewing the alternatives from the perspective of society as BCA does. It also explicitly considers the impact of uncertainty on the choice between alternatives through the construction of a decision tree, in which different branches reflect the likelihood of specific events occurring and their consequences as a function of the alternative being considered. It is thus relatively easy to examine the value of new information regarding climate change scenarios on the choice between alternatives. If one is interested in the ways that different interested parties will be affected by specific adaptation measures, then DA would be an appropriate approach to use. As with BCA, DA can be complemented by risk management as discussed above, but it is equally dependent on credible characterizations of the relative likelihoods of alternative futures.

Multicriteria Analysis

BCA and DA normally focus on a single metric by converting all impacts into monetary values. In many cases it may be appropriate to introduce multiple attributes into the analysis and examine the tradeoffs between them. For example, if there are environmental impacts that are difficult to convert into a monetary value or one wants to focus on the number of lives impacted by a particular alternative, then multicriteria analysis may be a more appropriate way to structure the problem. One way to combine these attributes would be to use a common metric such as utility and determine the weights that different attributes have in determining the aggregate utility of one alternative over another (see Keeney and Raiffa [1976] for techniques for doing this). Multicriteria analysis (MCA) could thus be used as a complement to BCA or DA in evaluating the relative attractiveness of different adaptation measures with respect to decisions at a societal level (BCA) or by specific interested parties (DA) (see Department of Communities and Local Government [2009] for an example of the use of MCA in the context of adaptation to climate change). Risk profiles borne of risk analysis and calibrated in whatever metric is most appropriate can also be employed to convert disparate vulnerabilities into comparable format (see, for example, Yohe, 2009b). In such cases, it is the political process that converts this information into relative rankings of significance for decision makers facing constrained resources (i.e., the political process then conducts its own MCA de facto).

Implementing Adaptation Plans

In order to move from planning to implementation, the following issues need to be addressed: prioritization of actions, establishment of time lines, availability of financial resources, and staffing needs. It is also important to establish a system to monitor the effectiveness of adaptations in achieving planning goals and to allow for adjustments to be made. Where appropriate, adaptation options that are not achieving their desired objectives should be adjusted, modified, or ended. In the latter case, it may be necessary to rethink how adaptations should be done. Box 4.7 summarizes how New York City addresses implementation in ways that recognize issues of timing (urgency) and expense.

IMPEDIMENTS TO IMPLEMENTING ADAPTATION PLANS AND POLICIES

The panel's evaluation of case studies has identified multiple barriers to effective implementation of adaptation programs and policies in the United States. Strategies

BOX 4.5
Characterizing Uncertainty in the Climate System

There are three main sources of uncertainty concerning future climate: the natural internal variability of the climate system, the trajectories of future emissions of GHGs and aerosols, and the response of the global climate system to any given set of future emissions (Meehl et al., 2007).

Internal variability refers to natural fluctuations of the climate system that occur in the absence of external radiative forcing due to, for example, increased concentrations in GHGs, aerosols from volcanic eruptions, or land use change. It is a result of the internal dynamics of the coupled atmosphere-ocean system. This internal variability includes natural fluctuations in large-scale phenomena such as the El Niño-Southern Oscillation.

Uncertainties regarding future emissions and concentrations of GHGs and aerosols are derived from imprecise understanding of exactly how the world will develop socially, politically, economically, and technologically. Uncertainties regarding future emissions of GHGs are often viewed as qualitatively different from the uncertainties associated with the physical climate system, and there is considerably greater controversy associated with quantifying uncertainties of emissions pathways since they involve quantifying uncertainties in very complex interactions of future world societies (CCSP, 2009c; Parson et al., 2007). It is generally viewed as unlikely that uncertainties of long-range emissions (e.g., more than a few years) can substantially be reduced (CCSP, 2009c).

Uncertainties about the response of the climate system to GHG emissions are normally analyzed using global climate models (GCMs). Different GCMs respond differently to the same radiative forcing and produce different patterns of climate change. These models provide information at relatively coarse spatial resolutions (hundreds of kilometers). The application of downscaling methods, such as regional climate models and statistical downscaling, yields higher-resolution projections but presents another source of uncertainty: the uncertainty associated with the spatial scale of the simulations. This last uncertainty can be particularly important in the context of adaptation planning, since adaptation studies or plans may require higher-resolution information about climate change.

Uncertainties that cannot be readily quantified through use of even the full suite of GCMs available throughout the world (see IPCC, 2007b) still must be recognized. These include such uncertainties as processes that are missing from the climate models (e.g., for some, a fully coupled carbon cycle, and evolution of land use or cover change), processes that are not explicitly resolved (e.g., deep convection) at typical global model spatial resolutions, processes that are not understood well enough to model successfully (e.g., certain aspects of ice sheet dynamics), and unknown processes. All of these uncertainties concern mainly the physical climate system. For further details on these physical system uncertainties, please see *ACC: Advancing the Science of Climate Change* (NRC, 2010b).

The three main uncertainties vary in their relative importance based on the prediction lead time of interest. For nearer time scales of one or two decades, internal variability dominates, whereas at longer time scales, model uncertainty and emissions scenario uncertainty dominate (Hawking and Sutton,

2009). The uncertainty in future human GHG emissions is the dominant contributor to uncertainty by the end of the 21st century.

Considerable effort has gone into developing means of quantifying the known uncertainties regarding future climate change on various temporal and spatial scales. These have included simple ranges of results from climate models (e.g., the range 2.7–8.1°F [1.5–4.5°C] for global temperature change response to doubling of carbon dioxide [CO_2]; Trenberth et al., 1995) qualitative statements of likelihoods (e.g., likely, very likely; Moss and Schneider, 2000), and probabilistic approaches (e.g., Webster et al., 2003; Wigley and Raper, 2001). The most widely used language of uncertainty is probability, and generation of probability distributions for various variables related to future climate has taken off in the past 10 years or so (CCSP, 2009c), for example, probabilities of climate sensitivity (e.g., Andronova and Schlesinger, 2001), probabilities of regional climate change conditioned on specific emissions scenarios (Tebaldi and Knutti, 2007), probabilities of emissions scenarios (Webster et al., 2003; Wigley and Raper, 2001), and probabilities of elements in society that contribute to emissions pathways such as future population (O'Neill, 2005).

Quantifying uncertainties in the physical climate system has relied primarily on two different types of climate model experiments: multimodel ensembles (MMEs), also known as ensembles of opportunity, which are made up of the results of different global climate models subjected to the same radiative forcing; and perturbed physics ensembles (PPEs), wherein key but uncertain parameters of a single global model are varied to essentially create a large number of different model versions (e.g., Murphy et al., 2004; Stainforth et al., 2005). Most recently, particular effort has been expended to quantify uncertainties of future climate on regional scales using probabilistic methods—sources of uncertainty that are perhaps more relevant to adaptation planning. These were particularly emphasized in the last report of the Intergovernmental Panel on Climate Change (IPCC) (Christensen et al., 2007). Most recently, the United Kingdom Climate Projections (UKCP09) has used a combination of PPEs, MMEs, and regional climate model results to develop probabilities of changes in temperature and precipitation at a 25-km resolution (Murphy et al., 2009) for all of Great Britain. It is expected that this information will be used to determine probabilities of different impacts of climate change and possible adaptations.

It must be remembered, however, that not all uncertainties can be easily quantified using straightforward probabilistic methods. Subjective probabilities have been recommended for establishing probabilities of different emissions pathways (Fisher et al., 2007), for example. The deep uncertainties that result from unknown or incompletely known processes may not be amenable to probabilistic quantification (Lempert et al., 2004).

Probabilities calculated for future regional climate change have in turn been used in impacts models (e.g., crop models and water resource models) to provide estimates of probabilities of impacts. This step can be viewed as part of a risk-assessment framework wherein the probabilities of the impacts form part of the input for risk assessment. One particular probabilistic approach that is directly relevant to adaptation planning is establishing probabilities of exceeding thresholds, for example, for some level of impact that may be beyond the coping range of a particular impact sector (Carter et al., 2007; Jones and Mearns, 2005).

BOX 4.6
The New York City Risk Matrix

In evaluating risk, New York City stakeholders filled in an automated template that measures risk as a factor of likelihood of impact, if a given climate hazard event should occur, and the magnitude of the consequence of such an impact. Depending on the response for each category, the spreadsheet then automatically generated a placement on a two-dimensional risk matrix (see figure below). If the automatically generated risk did not align with expert judgment, the spreadsheet provided the opportunity for override with notes explaining the override decision. If an adaptation measure was under way or planned and fully funded, stakeholders were instructed to take into account the benefits gained from those measures when conducting this exercise. Instances were also considered where updated measures not explicitly related to climate change adaptation were already under way within an organization or agency that would provide ancillary benefits for climate change adaptation.

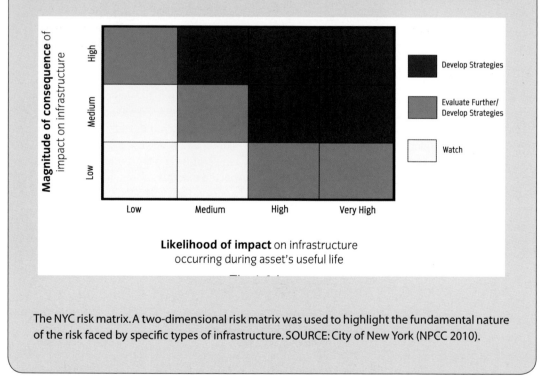

The NYC risk matrix. A two-dimensional risk matrix was used to highlight the fundamental nature of the risk faced by specific types of infrastructure. SOURCE: City of New York (NPCC 2010).

BOX 4.7
Casting Adaptation to Climate Change Within New York's Development Plans

The implementation step (number 7 in Figure 4.1) cannot be done in isolation. Investments in adaptation programs must be integrated into budget decisions that recognize a myriad of competing demands for scarce resources. In its planning process, New York City has concluded that it is essential that both the urgency and the cost of any proposed adaptation response be compared with other adaptation options so that its place in the long-term sustainability planning of the city can be determined and supported. The city developed a prioritization matrix to assist in this final step (see figure below). Notice, though, that the monitoring function in step 8 of Figure 4.1 must include not only adaptations that are implemented but also adaptations that are deferred. In that way, the flexible, iterative program can adjust its evaluation of urgency for the next round of decisions.

In summary, climate change planning was embraced in New York City as a way to integrate ongoing plans focused on growth management, infrastructure, and environmental sustainability. Climate change was chosen as the integrating element because adaptation to changes in climate-related risks could serve as a focal point. This "mainstreaming" of climate change planning into other ongoing initiatives moved adaptation to an advanced stage very quickly. The public-private initiative that was developed, with vigorous and effective leadership from the Mayor's Office, provided a coordinated approach while still allowing each stakeholder group to identify vulnerabilities and suggest pathways to resilience and sustainability.

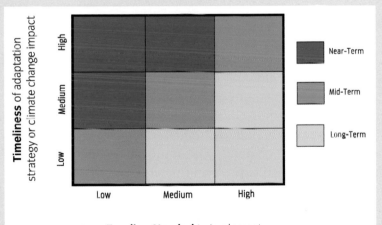

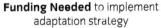

Prioritizing with respect to urgency and cost. Locating adaptation strategies that emerged from the evaluation process in a matrix that contrasts urgency with cost is the final step before an overall strategy for responding to climate change is brought into general budgeting and investment conversations. SOURCE: City of New York (NPCC 2010).

exist for overcoming each of these barriers, although some are more easily overcome than others:

Inadequate Information and Experience

- There is a high degree of uncertainty about future climate impacts at a scale necessary for most decision making, and there are uncertainties associated with the complexity of interactions between natural and socioeconomic systems. Because of these uncertainties, a variety of new management tools and region- and sector-specific data on likely climate impacts are needed, as well as new approaches to decision support.
- There is limited knowledge and experience with adaptation in the context of climate change among decision makers, resulting in a need for "learning by doing" and deliberately testing new approaches. The prospect of a nonstationary future climate system results in a need to design policies, monitoring activities, and processes that accommodate adaptive management.
- Limited stakeholder awareness of climate-related risks suggests that better information, public education, and decision support are required.

Inadequate Institutional Support for Adaptation

- Conflicting mandates within federal agencies and incentives for maladaptive behavior within governmental programs suggest a need to change regulations and incentives that currently increase climate-related risk.
- Lack of coordination across levels of government and between government and the private sector at multiple spatial scales means that mechanisms for interagency coordination need to be improved and networks of existing adaptation capacity will need to be strengthened.
- Inadequate institutional support for planning and implementing adaptations to climate change suggests that government and private-sector institutions need to be designed to (1) develop, interpret, and disseminate scientific information on climate change; (2) develop adaptation options; (3) help find resources to support adaptation programs; (4) enable adaptation projects to be implemented; and (5) monitor the success or failure of adaptations and enable corrective action to be taken. As with infrastructure-based solutions, however, institutions can either help or impede adaptations. Some institutions developed to support activities appropriate under prior climate conditions may discourage change and thus can impede adaptations.

Lack of Resources and Technology for Adaptation

- Many adaptation options will require financial investment, and insufficient capital or access to capital may limit options in many sectors or regions. If costs are too high, lower-cost options may be preferred or it may be possible to phase in adaptations to spread costs out over time. Inadequate resources can be addressed by (1) having a long-term investment strategy that includes identifying consistent sources of funding that are not subject to the vagaries of politics and (2) finding ways to limit the costs of adaptation, such as by mainstreaming adaptation into a wide range of decisions with climate-sensitive consequences, including reauthorization of laws affecting land and water use and construction or renovation of major infrastructure.
- New technologies may be needed to adapt to climate change. There are many types of technologies that can help adapt, including agricultural cultivars that are resistant to heat, drought, or excess moisture; water conservation technologies; and monitoring technologies. Research and development needs are further discussed in Chapter 7.
- The role of infrastructure in protecting existing ecosystems and valued investments needs to be considered. For example, having sufficient coastal protection or water storage infrastructure can help societies adapt to changing sea levels or diminishing water supplies.

Behavioral Impediments

- Behavioral impediments, such as a failure to acknowledge risks of climate change impacts, can be partially overcome through leadership, education, and better facilitation of decision processes. The likelihood that impacts will exceed the capacity to adapt in some regions and sectors creates a need to act quickly to reduce GHG emissions and to be better prepared for disasters.
- A short-term perspective among policy makers, which limits the capacity to address problems such as sea level rise, needs to be replaced by consideration of both short- and long-term benefits as well as multigenerational equity. There is a tendency to discount the future at a much higher rate than would be used in benefit-cost analyses (Loewenstein and Prelec, 1992). Furthermore, decision makers tend to ignore risks when perceived likelihoods fall below some threshold of concern (Huber et al., 1997).
- Underlying social and economic stresses that increase vulnerabilities to climate change impacts can be addressed by deliberately identifying and managing multiple stresses in an integrated manner. A sustainability-focused

solution set can be designed to solve multiple problems. Uneven access to adaptations among different income classes and populations can increase the vulnerability of society as a whole to climate change. The effect of inequity in vulnerability among such groups was clearly demonstrated in the immediate aftermath of Hurricane Katrina (Wilbanks and Sathaye, 2007).

It should be noted that in general, such barriers are far more prevalent in developing countries than in the United States (Smit et al., 2001). Therefore, the vulnerability of developing countries to climate change is generally considered to be much greater. This topic is considered separately in Chapter 7 (see also IPCC, 2007a).

LIMITS TO ADAPTATION

Potential adaptations to climate change have physical, economic, and institutional limits. For example, there are practical limits to how high seawalls and levees can be built, how much irrigation water can be applied, and how large storm sewers and culvert capacities can be. Institutional practice (custom, regulatory, legal) further constrains adaptations. Typically, major capital investments are financed over at most a few decades, whereas the infrastructure can last a century or more. A change in infrastructure design that will mostly provide benefits beyond the finance period may be difficult to justify (AWWA, 2009b). Furthermore, some adaptation options have maladaptive or unanticipated effects. A common example is the so-called levee effect, where establishing levees or seawalls encourages further development, thereby increasing catastrophic losses when the levees are eventually overtopped or seawalls breached (Tobin and Montz, 1997). A framework for considering thresholds for major "reasons for concern" was introduced in Chapter 2.

A thorough adaptation plan examines these limits to adaptation, even if probabilities of such outcomes are low (high-impact/low-probability events). As a result of considering such outcomes, the planning process might include evaluations of whether objectives may need to be significantly changed. This can be a politically challenging topic to address because it involves admitting that adaptation strategies may not or cannot meet the objectives under all conditions. Contingency plans for such outcomes should be developed, although much research is needed on how to develop such plans.

RESEARCH AND DEVELOPMENT IN SUPPORT OF ADAPTATION

While there are myriad adaptation options available to address vulnerabilities to current climate variability and extremes, research is urgently needed to devise innovative adaptation strategies to address the impacts of a changing climate. These strategies will need to tackle issues such as the unprecedented pace of change, the potential for crossing thresholds, the interaction of climate change with multiple other stressors, and the difficulties of anticipating the magnitude of extreme events. There is a major role for institutional innovation in adaptation. Box 4.8 suggests ways that the National Flood Insurance Program could be modified to reflect the risks associated with climate change while at the same time encouraging adaptation measures.

Among the areas where research can expand adaptation opportunities is in developing new technologies and management approaches and improving information on climate change, particularly at the regional and local scales where adaptation decisions will be made. Chapter 7 discusses in greater detail the major scientific and technological needs to promote effective adaptation to climate change. A key question decision makers might face is whether adoption and implementation of adaptations should wait for such improved regional-scale information. One option is for decision makers to develop robust adaptation approaches that will be effective in a range of future conditions, regardless of the pace of scientific and technological developments—especially since the degree of certainty about climate change trajectories that most decision makers desire is unlikely to be forthcoming.

Because of the considerable uncertainty regarding regional-scale climate change impacts, many stakeholders addressing climate change have expressed frustration with the lack of agreement on projections across global climate models as well as with these models' low spatial resolution. Research should aim to improve the resolution and accuracy of climate modeling and thereby increase confidence in the projections, particularly where consideration of adaptation options has identified specific informational needs. However, it is important to recognize that many adaptation decisions do not require precise forecasts of future conditions. For example, adaptations that incorporate flexibility, robustness (see discussion on robust decision making), or are hedging strategies may be effective under a wide variety of possible changes in climate (as well as under current climate conditions). The adoption of such adaptations need not await improved accuracy in climate predictions. Some decisions such as on infrastructure or other investments with a long lifetime can benefit from more precise climate projections. For such decisions, stakeholders need to weigh whether it is better to make a decision based on the current state of science or delay in the hope of having more precise projections in the future. Stakeholders should consider that while

BOX 4.8
Case Study in Innovative Adaptation: The National Flood Insurance Program

New mechanisms for public-private partnerships in adaptation measures are needed, particularly in reducing future losses from floods and hurricanes associated with climate change. Insurers provide private-sector financial protection to those at risk from potentially large catastrophic losses (e.g., due to earthquake, hurricanes, or terrorist attack) by charging a fee (premium) to those who seek such protection and, in turn, agreeing to pay all or a portion of the financial losses incurred in the event. Insurers that write policies for a large number of properties in a single geographical area face the possibility of large losses from a single event. The amount of coverage that an insurance company is willing to offer depends on the firm's capital management, regulatory approvals of rates, availability and price of risk transfer instruments, and the insurer's appetite for risk.

Insurance in the United States is regulated at the state level, with the principal authority residing with insurance commissioners. Insurance commissioners often regard solvency as a principal objective for insurers, even if it means requiring higher premiums or other insurer adjustments (e.g., reducing their catastrophe exposures). State governments also have created and operated catastrophe insurance programs following large-scale disasters to supplement private insurance and reinsurance.

Private- and public-sector insurers could play a role in encouraging adaptation to climate change risks in flood- and hurricane-prone areas.[1] This could be done by increasing insurance premiums to better reflect the value of the risk. Insurance premiums could be set at a price that captures the value of the asset as well as the level of risk present. Establishing premiums in this way would provide signals to individuals about the level of hazards they face and could encourage them to engage in cost-effective adaptation measures to reduce their vulnerability to catastrophes. Risk-based premiums could also reflect the cost that capital insurers need to integrate into their pricing to ensure adequate return to their investors.

The application of this approach would provide a clear signal of likely damage to those currently residing in areas subject to natural disasters and those who are considering moving into these regions. Risk-based premiums would also enable insurers to provide discounts to homeowners and businesses that invest in cost-effective loss-reduction actions. If insurance premiums are not risk-based, insurers have no economic incentive to offer such discounts. In fact, they prefer not to offer coverage to these property owners at all because it is a losing proposition in the long run.

In the context of the National Flood Insurance Program, applying this approach would require more accurate flood maps than currently exist. A recent National Research Council report (2009b) highlights the need for the Federal Emergency Management Agency (FEMA) to collaborate with federal, state, and local government agencies in this regard. The risks associated with losses from hurricanes and flooding may be higher than current estimates if there is an increase in the intensity of hurricanes or a higher-than-anticipated sea level rise caused by climate change during the next 10 or 20 years.

Pursuing risk-based premiums could dissuade development of hazard-prone areas, but it could have high, unexpected costs for those already living in these locations. The value of land in high-hazard zones has a wide range. It may be desirable, beachfront property prone to erosion; but it could be land that is affordable because it is vulnerable to flooding. Therefore, to address issues of equity and afford-ability, financing for increased premiums could come through general public funding for homeowners currently residing in hazard-prone areas, particularly low-income uninsured or inadequately insured homeowners, rather than through insurance premium subsidies.

The drawback of this provision is that it could discourage adaptation among residents of hazard-prone areas. As discussed in the next section, regulations imposed by state insurance commissioners keep premiums in many hurricane-prone regions artificially lower than the risk-based level, encouraging maladaptive behavior. If residents in these areas were provided with financial assistance from public sources to purchase insurance, it could directly encourage development in hazard-prone areas and exacerbate the potential for catastrophic losses from future disasters.

The complexity involved in adjusting insurance premiums, addressing concerns about equity, and accounting for behavioral biases suggests that innovative strategies are needed to encourage individuals to adopt cost-effective loss-reduction measures. Two possible complementary measures for dealing with this problem could be long-term contracts and well-enforced building codes.

Long Term Contracts for Encouraging Adaptation

Two types of long-term contracts could encourage individuals to invest in adaptation measures: long-term flood insurance and long-term loans. Today, flood insurance is only offered as an annual contract, and many property owners would be reluctant to incur the costs even if they received a premium discount the next year. If property owners underweight the future, or only focus on the expected reduction in losses for the next several years, they would not want to incur the up-front cost of an adaptation measure. An alternative approach, then, would be a 20-year flood insurance policy that would tie the contract to the property rather than to the individual.

Long-Term Home Improvement Loans

The second suggested measure would provide long-term home improvement loans, tied to the home's mortgage, for reducing vulnerability to climate change impacts. Such loans could be incorporated as part of the mortgage at a lower interest rate. A commercial bank would have a financial incentive to provide this type of loan: by linking the adaptation expenditures to the structure rather than to the current property owner, the annual payments would be lower, making the loan more attractive to mortgagees. The bank would be more fully protected against a catastrophic loss to the property, and FEMA's potential loss from a major flood would be reduced because of the investment in adaptation. These adaptation loans would constitute a new financial product, and the general public would see fewer tax dollars spent on disaster relief (Kunreuther and Michel-Kerjan, 2009).

---

[1]For a more detailed discussion of these principles see Kunreuther and Michel-Kerjan (2009).

the science of climate change projections can be expected to improve (e.g., as the models and techniques for analyzing multiple model projections improve), substantial enhancements in climate projection science may take many years or decades to be realized. Consequently, delaying adaptations in anticipation of improved science can increase risk, for example, by delaying taking measures to reduce exposure to climate hazards (Barsugli et al., 2009). Therefore, it remains necessary to evaluate the tradeoffs between acting based on the current state of science versus delaying in hope of having improvements (*ACC: Informing an Effective Response to Climate Change*, NRC 2010a; Barsugli et al., 2009).

CONCLUSIONS

Adaptation to the inevitable impacts of a changing climate has emerged as a pressing concern at all levels of government, but actions are still hampered by lack of engagement, lack of resources, lack of adaptation-related research, poor understanding of vulnerability, and limited capacity to improve these conditions in the face of competing policy priorities. Recent developments—the emergence of scientific consensus on causes and long-term trends in climate change; evidence that climate change impacts are already under way; realization that greater changes are coming, even if their timing and magnitude remain uncertain; and recognition that the past is no longer a reliable guide for the future—have validated the need for adaptation planning to manage risk. The existing barriers to adaptation illustrate that, without a well-integrated, comprehensive planning process and an adaptive risk-management approach, the United States is ill prepared at this time to efficiently and effectively deal with climate change impacts.

 Clear strategies and coordination across agencies and all scales of government within the United States will be essential to leverage limited resources; avoid redundant or conflicting projects, mandates, and guidelines; improve understanding of changing conditions; overcome behavior-based limitations to the capacity to adapt; and encourage learning as part of the policy-making process. Many states and municipalities have developed strategies and plans to adapt to climate change. These experiences have provided valuable insight into effective planning for adaptation, contributed to building the nation's adaptive capacity, and identified some win-win solutions to reduce the impacts of climate change.

In general, risk-management approaches to adaptation planning and action do not require a high level of precision about longer-term impacts of climate change, because they seek robust responses to a range of possible risks over time. Due to uncertainties about future impacts and contexts, risk-management approaches can assist planning and decision making because such approaches tend to emphasize options that offer co-benefits, that is, that have benefits for reducing sustainable development stresses as well as for improving the ability to cope with climate change.

Conclusion: A risk-management approach provides an appropriate framework for assessing the costs and benefits of adaptation options and prioritizing adaptation activities.

Conclusion: Governments, individuals, and organizations that are or may be affected by climate change need to begin to adapt by assessing their current and future vulnerabilities and developing adaptation strategies and plans. New York City provides an excellent example of ways to structure a public process for setting adaptation priorities and of the vital role of leadership in managing such processes.

Conclusion: An adaptation strategy needs to define a clear set of objectives that are focused on building adaptive capacity and reducing risk over multiple time frames, sectors, and scales. It needs to engage a wide range of participants, including those who are vulnerable and those who will be responsible for implementing the strategy. The success of the strategy will at least in part depend on the ability to engage both public and private sectors, to provide incentives for adaptive behavior, to communicate both risks and opportunities, and to prepare for gradual changes as well as low-probability/high-impact extreme events.

Conclusion: An adaptation plan to implement this strategy needs to include the following:

- An assessment of vulnerabilities in the context of other existing stresses, and the identification of adaptation options that are consistent with achieving broad societal objectives, including economic and environmental sustainability goals.
- A consistent methodology to analyze and evaluate adaptation options, including opportunities to "mainstream" adaptations within existing programs and processes and eliminate existing incentives for "maladaptive" behaviors. This should include consideration of the following:
    o Benefits (effectiveness) of adaptation, including reducing vulnerability to climate change impacts;
    o Co-benefits: positive effects on other systems or sectors;
    o Adverse impacts: negative effects on other systems or sectors;

o   Costs to implement the adaptations;

o   Overcoming barriers to adoption of the adaptations; and

o   Limits to adaptation: At what magnitude of climate change would the adaptations become ineffective?

- A plan for monitoring and evaluation of the adaptations in order to facilitate adaptive management of risks, learn from experience, and build adaptive capacity.

CHAPTER FIVE

Linking Adaptation Efforts Across the Nation

Adaptation is a process and not mainly about a set of actions to be taken right now. Nevertheless, this report does identify some "low-hanging fruit": near-term options to mainstream adaptation into current policies and programs (Chapter 8). Adaptation is primarily about developing a multiparty, public-private national framework for becoming more adaptable over time: improving information systems for telling us what is happening, both with climate change impacts and with adaptation experiences; working together across institutional and social boundaries to combine what each party does best; and making it a part of our national culture to continually revisit what risk management strategies make sense as we learn more about what we are facing from climate change.

In this sense, adaptation poses enormous challenges across sectors, jurisdictions, and levels of governance. Successful adaptation to climate change involves a multitude of interested partners and decision makers: federal, state, and local governments; the private sector, large and small; nongovernmental organizations (NGOs) and community groups; and others. The issue is how to create a framework in which all of the parties can work together effectively, taking advantage of the strengths of each and ensuring that their activities reinforce each other rather than getting in each other's way.

There are three general kinds of alternative approaches for meeting this need:

1. *A strong federal government adaptation program, nested in a body of federal government laws, regulations, and institutions.* With this approach, the federal government would take the lead in identifying adaptation actions in the national interest, mandate appropriate responses while providing resources to support them, set goals for improvements in the nation's adaptive capacities, and ensure coordination with other national programs and other parties nationwide.
2. *A grassroots-based, bottom-up approach that is very largely self-driven.* Adaptation planning and actions would be decentralized. Decisions would be made without significant federal encouragement or coordination, except for programs of the federal agencies themselves. Current adaptation efforts are largely occurring in this manner.

3. *An intermediate approach.* Planning and actions would be decentralized but the federal government would play a significant role as a catalyst and coordinator at the outset, providing information and technical resources, and continually evaluating needs for additional risk management at a national level.

In consultation with social scientists, practitioners, and stakeholders, this panel considered all three approaches and found that the intermediate approach had the strongest support among these groups. The examples presented in this chapter substantiate this finding.

ADAPTIVE CAPACITY

How do we build the capacity across the vast range of decision makers in government, businesses, and households throughout the nation to understand, assess, and address their vulnerabilities to climate change? As previous chapters have emphasized, vulnerabilities to climate change and options for adaptation are so diverse and so often specific to local contexts that adaptation decisions will need to be made and implemented by a wide variety of parties in all levels of government, business, and society at large. Vulnerability to the impacts of climate change depends on not just the exposure to impacts but also the sensitivity and the capacity to cope with the impacts. Therefore, assessing and building adaptive capacity will be a critical factor in determining the nation's vulnerability to the impacts of a changing climate (Adger and Vincent, 2005). In addressing this capacity-building challenge, there is often a mismatch between the scale at which an adaptation decision should be made and the capacities to adapt. This mismatch involves both knowledge about what to adapt to and the financial and human resources to make the adaptation happen. Because governance is an important determinant of adaptive capacity, this chapter's discussion focuses on the role of the public sector in adaptation and its capacity to carry out this responsibility (Finan and Nelson, 2009; Moser, 2009a).

> Adaptive capacity is the ability or potential of a system to respond successfully to climate variability and change, and includes adjustments in both behavior and in resources and technologies. (Adger et al., 2007)

The capacity to adapt to new stresses associated with climate change is uneven across and within sectors, regions, and countries (IPCC, 2007a; O'Brien et al., 2006). Although wealthy countries and regions have more resources to direct to this issue, the availability of financial resources is only one factor in determining adaptive capacity (Moss et al., 2001). Other factors include the ability to recognize the importance of the problem in the context of multiple stresses, to identify vulnerable sectors and communities, to translate scientific knowledge into action, and to implement projects and programs. The will of politicians (or the will of constituents to enable their representatives) to make decisions and spend resources on long-term investments is especially difficult given the uncertainties associated with the future magnitude of climate change; this critical ingredient of adaptive capacity is unfortunately often lacking. Furthermore, the capacity to adapt is dynamic and influenced by economic and natural resources, social networks, entitlements, institutions and governance, human resources, and technology (see Chapters 2-4 of this report; IPCC, 2007a). It is important to understand that nations with greater wealth are not necessarily less vulnerable to climate impacts, and to acknowledge that a socioeconomic system might be as vulnerable as its weakest link (Tol and Yohe, 2007). Therefore, even wealthy nations can be severely impacted by extreme events, socially as well as economically, as the United States learned from Hurricane Katrina (IPCC, 2007a). In fact, adaptive capacity itself involves diverse elements, which helps to explain why a certain response to a particular climate change impact can result in a positive outcome in one place but not in another (Tol and Yohe, 2007).

As described in Chapter 2, the United States is vulnerable to a wide range of climate change impacts such as increased droughts, sea level rise, flooding, loss of biodiversity, increased heat waves, and other effects. Although significant adaptation planning activities are already under way in some cities, states, sectors, NGOs, and federal agencies (Chapter 3), it is clear that there are many areas where adaptive capacity appears to be quite limited (see, e.g., Feldman and Kahan, 2007; Moser, 2009a,b; NRC, 2009a; and Chapter 4 of this report). Improving the nation's adaptation capacity requires, first, identifying the existing adaptation capacity within the private sector, NGOs, and state, local, federal, and tribal governments, and second, identifying gaps and high-priority adaptation actions that need greater resources and institutional support.

Because the nature of governance is an important determinant of adaptive capacity and, in turn, of the success of any adaptation to climate change (Finan and Nelson, 2009; Moser, 2009a), it is important to consider roles that the public sector can play in capacity building, along with the unique roles of the private sector and NGOs. Examples of public-sector roles include supporting adaptation plans and projects, monitoring climate impacts, reducing the vulnerability of infrastructure, providing information on risks for private and public investments and decision making, incentivizing investments in technologies and other adaptations that may have long-term benefits or cost savings, addressing needs for public education, building institutions and knowledge bases, and workforce development (Adger et al., 2007; NRC, 2007a). In addition, Chapter 6 describes the federal government's important role in considering the international context of adaptation to climate change impacts.

As Chapter 3 demonstrates, many response options will require cooperation or coordination across jurisdictional boundaries or between agencies with potentially conflicting goals. Consequently, efforts to reduce climate risks are more likely to be efficient and effective if such adaptation activities are well coordinated. This means that capacity building includes developing institutional frameworks for coordinating adaptation planning and actions across geographic scales, sectors, and categories of decision making.

The challenge of coordinating across scales illustrates this point. Processes and actions that shape both climate change impacts and adaptive responses interact constantly at scales from global to local, and these interactions can undermine effective adaptation. An overemphasis on top-down adaptive strategies may result in solutions that are insensitive to local contexts, a backlash from local stakeholders, and a lack of empowerment of local creativity. An overemphasis on bottom-up strategies may result in limited sensitivity to larger-scale driving forces, a limited understanding of spatial context across jurisdictional boundaries, and a lack of access to resources to support effective actions (Wilbanks and Sathaye, 2007).

ROLES OF GOVERNMENTAL AND OTHER INSTITUTIONS

Because the impacts of climate change affect a wide range of public and private interests and will likely require significant resources, governments will play a number of key roles in coordinating, supporting, and implementing adaptation measures in the United States. However, the complexities of climate change and the interactions among various impacts, affected stakeholders, and governing entities present a major challenge in creating effective institutional frameworks for adaptation.

An important rationale for government engagement in the design of adaptation strategies and plans is that responses to climate change often need to take place within a comprehensive process that deals with multiple stresses threatening vulnerable populations, resources, and systems, many of which involve government roles. For example, natural resources managed primarily by government and under jurisdiction of local, state, and federal governments require comprehensive planning efforts to address the impacts of climate change and other pressures (West et al., 2009). In addition, as discussed in Chapter 2, populations at risk from climate impacts are often already at risk due to age, poverty, lack of access to services, or other stresses. Increased exposure to climate impacts over time (heat waves, sea level rise, hurricanes, extreme droughts, wildfires, extreme precipitation events, etc.) will exacerbate these risks. Indeed, the Intergovernmental Panel on Climate Change (IPCC, 2007c) makes it clear that (1) the poor, the sick, the elderly, and the young (especially those living in megacities near the coast) are most vulnerable to climate change and (2) communities facing compounding risks from multiple stresses can be found everywhere—even in the wealthiest countries on Earth. It follows that much of the underlying vulnerability to climate change will not be ameliorated by climate-specific adaptation programs that are constructed without acknowledging multiple stresses, and that the interjurisdictional nature of these problems requires the engagement of government at multiple levels. Therefore, it will be important that comprehensive planning occurs not only within affected sectors and populations but also across and between those parties at local, state, regional, and national scales.

Roles of Local Governments

Many adaptation decisions and actions are being and will be made by governments at a local scale. Under the powers granted to the states by the Tenth Amendment to the U.S. Constitution, states often delegate broad authorities to local governments for comprehensive planning and land use controls to protect the health, welfare, and general safety of their citizens (Porter, 1997). Local governments also fund and make key decisions regarding public infrastructures including water, solid-waste, wastewater, and stormwater systems; transportation; natural resources; and public facilities such as schools, hospitals, and public housing. Local governments are also responsible for emergency preparedness, response, and other aspects of public health and safety. However, local emergency preparedness plans rarely directly address the impacts of climate change. Likewise, local land use regulations usually do not address anticipated sea level rise, increased storm surges, or changes in 100-year flood cycles, and many other local systems are currently failing to consider climate change vulnerabilities.

Local governments are increasingly recognizing and addressing climate change adaptation through various planning efforts (examples in Chapters 3 and 4 in this report; Feldman and Kahan, 2007; a recent inventory is provided by Moser, 2009a). For example, the International Council on Local Environmental Initiatives (ICLEI, recently renamed Local Governments for Sustainability), with funding from the National Oceanic and Atmospheric Administration (NOAA), launched an initiative in 2005 to assist local governments in conducting vulnerability assessments and improving "resiliency" to climate change impacts. Initial partners in this initiative included Keene, New Hampshire (see Box 5.1); Fort Collins, Colorado; Anchorage, Alaska; and Miami-Dade County, Florida. ICLEI has since worked with King County, Washington, in the development of a guidebook for local governments in preparing climate adaptation plans (Snover, 2007). King County was one of the earliest leaders among local governments in comprehen-

BOX 5.1
The Case of Keene, New Hampshire

Keene, New Hampshire, a city of 23,000 that is one of five pilot communities in ICLEI's new Climate Resilient Communities program, signed on to a climate change effort in 2000 to reduce its greenhouse gas (GHG) emissions to 10 percent below 1995 levels by 2015. Keene had already begun to experience more intense rainfalls, a major 500-year flood in 2005, decreases in snow days, infestations of nonnative plant and animal species, and more days with high heat and poor air quality. All of these contributed to willingness to undertake the pilot effort to become a climate resilient community. This resulted in the formation of a committee to identify climate change impacts, community vulnerabilities, and opportunities for adaptation and mitigation, and to establish goals and targets to achieve resilience. Much of the success of the committee reflected active participation by the mayor, city manager, department heads, City Council members, the local Climate Protection Committee, college faculty, regional planners, and public health responders.

Another contributing factor to the successful launch of this effort was the assistance Keene received from NGOs, boundary organizations (see Box 5.5 for definition), and the federal government. ICLEI-Local Governments for Sustainability (ICLEI), an international NGO, provided Keene with a sense of affiliation with a larger movement, staffing for the pilot study, a tested template of five key milestones used previously in limiting GHG emissions, and techniques for identifying vulnerabilities, choosing targets and priorities for adaptation, and sharing the experience nationally. NOAA supported the work through its Regional Integrated Sciences and Assessment (RISA) staff, using Environmental Protection Agency (EPA) data and a major study of the Northeast by the Union of Concerned Scientists that provided details of New Hampshire climate impacts, as well as studies by faculty at Antioch University and the University of New Hampshire. Overall, the ICLEI approach and template provide a significant strategy for designing an effective, action-oriented adaptation plan based on local government, but one that needs improved regional climate impact data—which demonstrates that local efforts depend on national and regional support.

As the Keene committee identified vulnerable sectors, they grappled with their inability at times to identify actions for addressing these problems because the ICLEI "priority template" was too general to make meaningful choices. They also found it difficult to understand how to distinguish adaptation actions from GHG-reduction measures, and climate-related actions from general sustainability and green-economy issues. To the Keene participants, GHG-reduction measures represented an effective form of adaptation, and climate change responses were part of a larger need for sustainability.

In the year and a half since publishing its plan, Keene officials have undertaken a number of actions to increase resilience, including investigating improved building design standards to withstand expected climate change impacts and the use of wetlands for flood storage. Their major effort, however, is to make the goals and targets of the adaptation plan part of the everyday process of local development, permitting, and code enforcement by including climate change adaptation, mitigation, and sustainability in the Community Master Plan to be approved in 2010.

sively addressing climate adaptation. In 2006, the county formed an interdepartmental climate change adaptation team and has already begun implementing a number of adaptation efforts (Cruce, 2009).

The New York City case study in Chapter 4 provides another example of leadership by a local government (albeit a very large one) resulting in a comprehensive planning effort. It also illustrates the unique challenges facing large urban centers. The United States currently has 29 cities with more than 500,000 residents, and all are likely to face significant challenges from climate change that involve infrastructure investments, protection of natural resources, vulnerable populations, and emergency preparedness. Under existing circumstances, adaptation to climate change could negatively affect the financial viability of some U.S. cities (with special concern for coastal cities; see Box 5.2) by increasing infrastructure maintenance and capital improvement project budgets, while at the same time potentially decreasing operating revenues through real estate devaluation. For example, climate change may require cities to write off investments in inundated or damaged infrastructure (while in many cases continuing to pay debt service on those lost assets), invest in new replacement infrastructure despite stressed financial resources, increase operating budgets to maintain service levels, and absorb potential drops in their property tax base.

Inadequate planning and adaptation choices could result in a significant downgrade of a city's bond rating by rating agencies, which would limit future borrowing potential and increase debt costs. In short, cities might need federal assistance or increased taxing authority to be able to finance new infrastructure or pay the operating costs associated with adjustments to climate change.

Despite strong leadership in initiating planning efforts in many cities, local governments agree that more support is needed for a nationwide response (U.S. Conference of Mayors, 2008). An important factor in adaptive capacity is the availability of technical and human resources to identify vulnerabilities, and the knowledge to make effective adaptation decisions (NRC, 2009b). Because of the current deficit in the knowledge needed to guide adaptation decisions at all levels of governance, the U.S. Conference of Mayors passed a resolution concerning "Climate Change Adaptation and Vulnerability Assessments" (U.S. Conference of Mayors, 2008), which called for Congress and the federal government to

> Pass climate change adaptation legislation that provides:
> * incentives to state and local governments to begin exploring the growing risks from climate change, conduct climate vulnerability assessments that identify

BOX 5.2
Roles of Federal, State, and Local Governments: Honolulu Case Study

As sea level rises, many coastal cities face a daunting adaptation challenge. Honolulu is one of the nation's larger coastal cities, and its adaptation needs are illustrative of the problems that these cities face. Moderate increases in sea level have the potential to inundate the Honolulu International Airport; the city's central Sand Island sewage treatment plant; thousands of coastal residential units; water, wastewater, and transportation infrastructure; as well as the Waikiki peninsula, the economic engine for tourism in the State of Hawaii.

For Honolulu to effectively adapt to these challenges, it will require the coordinated efforts of federal, state, and city governments. For example, since the state owns the airport and the harbors, state government will need to work with city leaders on relocating or protecting these facilities in place and to support the redevelopment of city infrastructure that services them.

The city might decide to alter its land use plans to relocate displaced coastal populations. If so, it will also need to reroute transportation corridors and redesign and rebuild water and wastewater transmission systems, pump stations, and treatment plants that are projected to be inundated. The city government, by itself, does not have the technical expertise to determine how, when, and to what degree sea level is likely to rise in their geographic area. This limits their ability to determine the amount, timing, or type of investments needed to prepare for sea level rise. Inadequate or misdirected infrastructure investments would leave the city exposed to coastal inundation. On the other hand, excessive investment in unneeded infrastructure improvements would also have negative financial consequences.

To limit these uncertainties, the federal government needs to provide the necessary guidelines—for example, by updating Federal Emergency Management Agency (FEMA) flood maps—to reflect future projected sea level rise in local jurisdictions. These updated maps can serve as guidance for city decision making on land use issues and infrastructure investment and also provide the city with the basis for the regulatory framework needed to direct future growth away from projected inundation areas. The federal government can also provide support to cities like Honolulu by establishing a uniform methodology to assess their vulnerabilities to climate change and to develop an adaptation plan. By utilizing a uniform methodology, the relative priorities of investments can be evaluated and the approximate capital and operating costs for various response options can be estimated. With such tools, cities like Honolulu can then make their own decisions about such things as abandoning or attempting to "armor" coastal areas. A uniform assessment methodology would also enable the federal government to make fair and reasoned decisions about how it allocates financial resources to affected cities.

the most important climate risks for a particular area or population, identify the response options, and ways to implement them; and

- assistance to state and local governments to develop climate change adaptation plans and to provide financial and technical assistance and training to state and local governments to implement those plans; and

- a national climate change adaptation strategy to combat adverse impacts of climate change to the economy and the environment and reduce the vulnerability of the nation's cities to the impacts of climate change and also urges the Federal Government to conduct annually national climate change vulnerability assessments; and

- methods and tools for studying climate change impacts on communities and integrating this information into state, regional, and local adaptation planning efforts.

Roles of States, Territories, and Commonwealths

States' responsibilities include natural resources, public health, emergency planning and response, public infrastructure, insurance markets, taxes, and managing state lands. The division of responsibilities between cities and states differs between regions; for example, some states exert direct authority over land use planning and regulation at the local scale, so the roles mentioned for cities above will be applicable to some states (Salsich and Tryniecki, 1998; So et al., 1986). States also provide technical assistance and funding for local projects and often coordinate emergency management activities and federal programs in support of local governments.

States have often become laboratories for new, innovative policies, and some have been described as taking the lead in climate adaptation planning in the United States (Feldman and Kahan, 2007; Moser, 2009b). While several states have cited sea level rise in the establishment of long-standing policies related to coastal erosion and inundation (for example, in the development of beachfront rules in South Carolina and Maine as early as 1987; see Moser, 2009b), a number of states have more recently (in the past 3 to 4 years) engaged in comprehensive climate adaptation planning. According to a recent survey by the Pew Commission on Global Climate Change, 8 states (Arizona, Colorado, Iowa, Michigan, North Carolina, South Carolina, Utah, and Vermont) currently recommend creating plans for adaptation in their climate action plans, and 10 states have begun comprehensive adaptation planning efforts that parallel ongoing planning activities for GHG emissions reductions (Alaska, California, Florida, Maryland, Massachusetts, New Hampshire, New York, Oregon, Virginia, and Washington; Pew Center on Global Climate Change, 2009; see also the California case study in Box 5.3, and the Alaska case study in Box 3.1). Many of these initial efforts have placed their strongest emphasis on additional research and monitoring needs rather than significant changes to state policies (Feldman and Kahan, 2007; Moser, 2009a). U.S. island territories and commonwealths are particularly vulnerable to coastal impacts and water shortages associated with climate change (USGCRP, 2009); however, adaptation planning efforts in the islands have only recently begun (see Tompkins et al., 2005).

The calls for research and monitoring in state plans confirm the conclusion of several previous National Research Council reports (NRC, 2007d, 2009a,b) that research and assessments undertaken as mandated by the U.S. Global Change Research Act (P.L. 101-606, 104 Stat. 3096-3104) have not yet produced the necessary information and decision-support tools to allow policy makers at the state level to initiate action on adaptation. Although the governor of California has recognized the need to develop a policy in response to climate change, the decision-relevant information on sea level rise and its socioeconomic implications for the state is lacking.

The need for decision-relevant information is also reflected in a recent resolution passed by the National Governors Association (NGA) calling for increased federal support of adaptation in relation to the coastal impacts of climate change. While focused on coastal issues, this resolution has broader implications for intergovernmental coordination of climate adaptation activities:

> Federal agencies are currently collecting useful data and administering programs for climate change adaptation, in addition to providing a range of federal funding sources to assist adaptation-related activities. Adequate intergovernmental coordination is needed to ensure the most effective implementation and efficient use of funds; provide opportunities for complementary efforts among local, state, regional, or national programs; and improve awareness and understanding of the resources available to states and local governments. Congress and the Administration should develop a national strategy to ensure intergovernmental coordination on coastal adaptation, clearly define the roles of various agencies, and identify the mechanisms by which federal programs will coordinate with state partners on coastal adaptation issues. (NGA, 2009)

In addition, governors urge Congress and the Administration to recognize the critical role of states in climate change adaptation policy by:

- ensuring consultation with states in any new legislation, programs, or research;
- developing a strategy to identify the information needs of states to effectively respond to natural hazards and ecosystem changes;

- coordinating any federal agency activities, research, and data collection efforts related to coastal impacts with states; and
- clarifying the roles and responsibilities of states and federal agencies in adaptation activities."

BOX 5.3
California Adaptation Plan: Case Study

In 2008, California Governor Arnold Schwarzenegger issued Executive Order S-13-08, calling for a detailed, statewide study of sea level rise implications, and for the California Natural Resources Agency to "coordinate with local, regional, state, and federal public and private entities to develop a state Climate Adaptation Strategy." In August 2009, the draft adaptation strategy was released for public comment It covered seven major topic areas, and each topic area was assigned a lead state agency or agencies: public health, biodiversity and habitat, ocean and coastal resources, water management, agriculture, forestry, and transportation and energy infrastructure.

The strategy followed the general principles to use the best available science, to design a flexible strategy recognizing that knowledge about climate change is still evolving, and to involve all relevant stakeholders throughout the process to establish and retain strong partnerships. Participating agencies are directed to seek adaptation strategies that contribute to social and environmental resilience and sustainability and build on existing policies rather than requiring new policies. Some of the resulting recommendations of the strategy included the following:

- State agencies should implement strategies to achieve a statewide 20 percent reduction in per capita water use by 2020.
- New development should be prevented in areas that cannot be adequately protected from flooding due to climate change.
- State agencies responsible for public health, infrastructure, or habitat subject to significant climate change impacts should prepare agency-specific adaptation plans, guidance, or criteria by September 2010.
- State agencies should identify key land and aquatic habitats at risk and develop a plan for expanding protected areas or altering land and water management practices.
- Communities with local coastal plans or general plans should amend them to assess climate change impacts and vulnerabilities and develop risk-reduction strategies.
- State firefighting agencies should begin immediately to use climate change impact information to inform future fire program planning efforts.
- Existing and planned climate change research should be used for state planning and public outreach purposes; new climate change impact research should be broadened and funded.

Roles of the Federal Government

The federal government has overarching responsibilities for natural resources, public health, emergency planning and response, taxes, and the management of federal lands, and also protects the national interest, national security, and homeland security. (The federal government's role in international relations, development assistance, treaty negotiations, and transboundary issues is discussed in Chapter 6.) For climate adaptation, the federal government has several roles to play, including

- Addressing transboundary and interjurisdictional conflicts, resources, vulnerabilities, and adaptations;
- Benefiting from economies of scale and established federal capacities in scientific and technical research and training on adaptation;
- Protecting assets necessary for functions historically assigned to the federal government, including securing interstate commerce and providing for national security;
- Protecting existing federal infrastructure investments, including highways, sewage and water treatment plants, ports, dams, and other infrastructure threatened by climate change;
- Protecting federal lands; and
- Ensuring adaptive risk management in other federal agency facilities and programs.

The federal government can also play a key role in supporting adaptation at regional, state, and local scales by providing coordination, guidance, and financial and technical assistance in response to user needs.

Due in large part to the U.S. Global Climate Research Act of 1990, the federal government is currently more heavily invested in climate modeling, monitoring, and mapping activities than are state and local governments. The U.S. Global Change Research Program supports federally coordinated research on global climate change and periodic assessment of the impacts of these changes on the natural environment, biological diversity, various sectors, human health and welfare, and social systems. While this research program has advanced understanding of the drivers of climate change and climate change impacts, it has not been designed to provide the information needed at regional, state, and local scales to address the impacts of climate change (NRC, 2009a,b). Therefore, organizations like the Conference of Mayors and the NGA, as well as several previous studies, have drawn attention to the need for the federal climate change research effort to increase its emphasis on providing more effective decision support in order to enhance the nation's capacity to adapt to climate change (Feldman et al., 2008; NRC, 2007a, 2009a,b).

Adaptation to climate change is beginning to occur at all scales, as this report demonstrates. The federal government could greatly facilitate the ongoing process of adaptation by supporting innovation, providing incentives for adaptation, and providing a network or clearinghouse through which successful examples of adaptation can be shared (GAO, 2009; NRC, 2009c). Congress has taken up the issue of climate adaptation in recent years, but it has yet to pass broad adaptation legislation (Feldman and Jensen, 2008; Moser, 2009b). A number of federal agencies have begun to independently address climate change adaptation options, most notably those involved in managing natural resources, federal lands, and transportation (see, e.g., CCSP, 2008a,b, 2009b; DOI, 2009; EPA, 2009; Fagre et al., 2009). However, beyond efforts recently initiated within the Office of Science and Technology Policy and the Council on Environmental Quality, there Is currently no comprehensive adaptation strategy, guidance, or coordination effort at the national level (GAO, 2009b). Consequently, stakeholders not represented by these federal agencies engaged in adaptation planning may not have equal access to federal support in capacity building (NRC, 2009c).

As with initial state efforts, most of the federal activities related to adaptation have focused on research and information needs or the identification of policy options rather than the establishment of significant new adaptation policies within federal programs (Feldman and Kahan, 2007; Moser, 2009b). In many cases, federal agencies have faced significant hurdles or other obstacles in modifying policies or regulations to address climate change. There is currently a shortage of federal leadership, funding, awareness, and coordination in considering both major constraints and competing priorities, lack of mandates, and legal obstacles—particularly rules and regulations that are based on historical patterns and fail to recognize that such patterns do not reflect emerging changes in climate and their impacts (GAO, 2009b; Moser, 2009b). For example, the Fderal Emergency Management Agency (FEMA) cannot require the use of future sea level rise projections for floodplain management or insurance ratings unless statutory and regulatory changes are made to the National Flood Insurance Program (NFIP; CCSP, 2009b; ASFPM, 2007). Other issues surrounding the NFIP are described in Box 4.8 in Chapter 4.

Because of the complexity of the task and the importance of consistency across sectors and federal agencies, federal leadership will sometimes be required to provide the necessary framework and strategy to assist decision makers (GAO, 2009b). Through interviews with governmental officials at local, state, and federal levels, the U.S. Government Accountability Office (GAO) report finds that some federal policies, programs, and practices can hinder adaptation efforts. In addition to citing issues with FEMA's floodplain maps, as mentioned above, respondents to the GAO questionnaire also criticized the U.S. Department of Agriculture Federal Crop Insurance Corporation for failing to take into account potential changes in the frequency and severity of weather-related impacts. Conflicts or limitations related to other federal mandates, such as the Endangered Species Act, the Clean Water Act, and the Clean Air Act, were also cited as potentially constraining options for resource management measures targeted for adaptation. Often, existing mandates were crafted with the intent of maintaining the status quo or returning environmental systems to some prior condition; these goals may be unworkable under a changing climate.

These statements echo previous and more general GAO findings (GAO, 2005a,b) regarding conflict among federal mandates and agencies:

> Agency missions may not be mutually reinforcing or may even conflict with each other, making consensus on strategies and priorities difficult. Incompatible procedures, processes, data, and computer systems also hinder collaboration. The resulting patchwork of programs and actions can waste scarce funds and limit the overall effectiveness of the federal effort. In addition, many federal programs were designed decades ago to address earlier challenges, informed by the conditions, technologies, management models, and organizational structures of past eras. (GAO, 2005a)

> Based on our prior work, key practices that can help agencies enhance and sustain their collaborative efforts include (GAO, 2005b)

> - defining and articulating a common outcome;
> - agreeing on roles and responsibilities;
> - establishing compatible policies, procedures, and other means to operate across agency boundaries;
> - identifying and addressing needs by leveraging resources; and
> - developing mechanisms to monitor, evaluate, and report on results.

The importance of governance and institution building at a national scale to capacity building and successful adaptation has been recognized by many foreign nations, most notably Australia (see Box 5.4), the United Kingdom (see Box 5.5), and Bangladesh (Box 6.2). In Australia, climate change is ranked as the greatest threat to Australia's national interest by 84 percent of people surveyed, and tackling climate change is overwhelmingly seen as important (90 percent; Howden, 2009). This general perception is reflected in the institutional change that occurred to respond to climate change (Figure 5.1). In Australia, it is recognized that adaptation is fundamentally a social process, which requires the integration of science and policy in a fundamental

BOX 5.4
Adaptation of Water Rights in Australia to Facilitate Markets and
Protect Environmental Flows

Chronic concerns about the deterioration of river flows resulted in a turning point in Australia's water policy in 1994, when the Council of Australian Governments agreed to a water reform framework that formally recognized environmental water uses as a legitimate water right. Another key feature of the 1994 reforms included the introduction of a cap on diverting water from the Murray Darling Basin,[1] the area with the largest agricultural water use in Australia. The cap became permanent for the states of New South Wales, Victoria, and South Australia in 1997, limiting extractive use to the rate in 1993-1994.

The cap on diversions and the provision of water for the environment have been accompanied by the separation of water access entitlements from land titles and the establishment of a market for water trades. The cap's main purpose is to keep water use and development within an agreed limit and maintain a degree of supply reliability at the individual level. To access more water than is currently allocated to a water right, someone else must agree to take less water. In rural areas, this is achieved primarily by letting people trade water on a temporary or permanent basis (Young and McColl, 2009).

Another step change in precipitation in Australia occurred around 1997, when average water inflows had fallen by roughly 50 percent in reservoirs that serve major coastal cities. Southeastern Australia, in particular, has been facing substantial water supply shortages requiring dramatic changes in water-use patterns. Water use for agriculture is now roughly one-third of what it was prior to 1997. In response to worsening conditions, additional water rights reforms were initiated through the National Water Initiative of 2004. Water rights are now divided into two components, a right to a proportional share of water (which includes a proportion for the environment), and an annual allocation. The annual allocation is the specific volume of water allocated to rightholders in a given season. Like a share in a public company, the return on a water entitlement has an average yield and is associated with a level of reliability. For example, a 100 ML entitlement may have an average yield of 90 percent (i.e., 90 ML). This approach to quantifying water rights allows the annual water rights volumes to fluctuate based on water supply availability.

---

[1]The Murray Darling Basin is located in southeastern Australia and extends across the jurisdictional boundaries of New South Wales, Victoria, South Australia, Queensland, and the Australian Capital Territory.

and structural way. A national fund was established to incentivize adaptive behavior and innovation within states. In addition, the Australian version of the National Science Foundation was reorganized as an interdisciplinary program to support research on system solutions and applied research. This is a deliberate attempt to "mainstream" climate change issues into policies and programs in general.

BOX 5.5
United Kingdom Climate Impacts Programme

The United Kingdom Climate Impacts Programme (UKCIP) is defined by stakeholder needs. The motivation from the beginning has been engagement: the users help to pay for the impact research and are partners in the projects.

UKCIP is a "boundary organization" that is intended to bridge the gap between climate science and society. It is funded through a contract between the government and the School of Geography at Oxford University. UKCIP defines the conditions under which they are willing to engage; it is a strong advocate of practical approaches, and learning by doing. Participants include flood, water supply, energy and environment agencies, water companies, sewage disposal companies, and a marine national park that uses climate information. These are examples of customers for which the UKCIP builds prototype products with hope for broader impact. For example, it produced an "adaptation wizard" tool with versions 1, 2, and 3 that takes into account probabilistic information and helped local authorities to use the tool. After UKCIP educates an early adopter, it encourages the transfer of the practice from one local authority to another, which is a good method of dissemination and builds on the lessons learned by the initial adopter.

A large investment has gone into downscaling regional climate change impacts and helping people understand the probabilities of different outcomes. "UKCP09"[1] provides the latest information on how continued emissions of GHGs may change the United Kingdom's climate over the 21st century. UKCP09 uses probabilistic projections at a resolution of 25-km (approximately 15.5-mile) grid squares for seven overlapping 30-year time slices to 2099. It also includes information for administrative and river basin areas. The higher spatial and temporal resolutions make these scenarios particularly useful for planning at the local level (UKCIP, 2009).

UKCIP has found that stakeholders need help defining what the right questions are at the regional level and that a way is needed to connect those questions to the national and international science network. Identifying stakeholder science needs, translating these science needs to the science community, and helping to translate science into relevant decision-support information is an iterative process, which is facilitated by the UKCIP. Its long-term funding and stability as a boundary organization has been critical to its success and has allowed the organization to develop trust with its stakeholders for more than a decade. The institution's mission is to start adapting despite the gap in scientific information and to reduce the existing adaptation deficit.

SOURCE: Chris West, UKCIP.

---

[1] See *http://ukclimateprojections.defra.gov.uk/* (UKCIP, 2009).

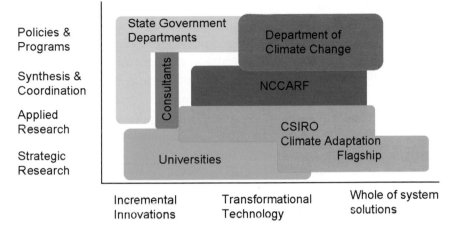

FIGURE 5.1 Diagram illustrating the new institutional landscape in Australia to facilitate collaboration on climate change adaptation. NCCARF, National Climate Change Adaptation Research Facility; CSIRO, Commonwealth Scientific and Industrial Research Organization. SOURCE: CSIRO Climate Adaptation Flagship, Australia, 2010.

In addition to its Climate Adaptation Flagship, Australia's federal government assumed central leadership in addressing climate change challenges by creating the Department of Climate Change and Energy Efficiency to coordinate the nation's climate change response based on three integral pillars: mitigation (i.e., limiting the magnitude of future climate change), adaptation, and international engagement (Commonwealth of Australia, 2009). The department develops the nation's climate strategy and has the overall mission to reshape the nation's economy and social system to align with the department's goal of reducing GHG emissions and adapting to the impacts of climate change.

The commitments at the national level by the United Kingdom and Australia have increased adaptive capacity by providing vulnerability and impact assessments, tools, and technologies required to implement adaptation. These efforts have led to some successful adaptation actions by the agricultural and water sectors in Australia (see Box 5.4 for additional details).

Roles of Regional Institutions and Boundary Organizations

Many climate impacts and responses do not conform to existing political or jurisdictional boundaries. Lessons available from long-standing natural resource management programs indicate that the most effective management programs are focused on the geographic boundaries of the resource itself, such as a watershed (for example, see Federal Agencies, 2000). Both rivers and aquifers cross state lines, and there are many cases of interstate stream commissions and special authorities to manage interstate and international watersheds. Likewise, resource issues such as wildfire, drought, invasive species, and flooding are interconnected and driven by large-scale climate processes, requiring coordinated and integrated preparedness plans. Addressing water, ecosystems, and other regional resource issues requires management responses across sectors and levels of government. Illustrations of the importance of cross-boundary issues are found in the case of managing endangered species in the Lower Colorado River watershed (see Chapter 6) and in the recent changes in water property rights in Australia. The Australian example illustrates both the need for drastic changes from business as usual and the need for national leadership to address some of these draconian measures (see Box 5.4).

Some states are also addressing sea level rise and related impacts through regional partnerships. For example, according to the West Coast Governors Association agreement, "[t]he West Coast states will focus initial efforts, in collaboration with the federal government, on a West Coast-wide assessment of shoreline changes and anticipated impacts to coastal areas and communities due to climate change over the next several decades, and work together to develop actions to mitigate and adapt to the impacts of climate change and related coastal hazards" (WCGA, 2008). On the East Coast, the Action Plan for the Northeast Regional Ocean Council (NROC) is also seeking to "Render New England a Coastal Hazards Ready Region" by coordinating interstate planning and data acquisition strategies (NROC, 2009). To the degree that such regional approaches define solutions at the scale of the problem, they are likely to be more effective than fragmented approaches across political boundaries.

Adaptation efforts can also benefit from "boundary organizations" that link climate science and technology with local decision makers to strengthen adaptive capacity. For example, NOAA's Regional Integrated Science and Assessment (RISA) program consists of nine teams operating within and serving different regions of the United States. These teams have developed innovative place-based, stakeholder-driven research, partnership, and services programs with a primary objective of improving adaptive capacity. RISA teams are comprised of researchers from the physical, natural, and social sciences as well as the fields of economics, geography, engineering, and law who work together and partner with stakeholders in a region to determine how climate impacts key resources and how climate information and tools could aid in decision making and planning for those stakeholders. RISAs are viewed as exemplars of place-based research that focuses on addressing specialized needs for climate information within regions and sectors, in part because they support long-term relationships between research teams and stakeholders that allow for collaborative learning (NOAA, 2009). Another effective model of a climate adaptation boundary organization is found in the United Kingdom Climate Impacts Programme (see Box 5.5).

Roles of the Private Sector and Nongovernmental Organizations

Effective, proactive adaptation will require participation of every category of decision maker, including business and industry, professional organizations, environmental groups, NGOs, social service and health organizations, and the research community. Adaptations implemented by government alone will fail to meet needs of many parts of U.S. society, will fail to take advantage of knowledge and capacities of institutions and parties outside of government (as illustrated in Box 5.1), and will miss the important fact that most adaptations to impacts of climate change will be made voluntarily by parties across the country, responding to information and experience, without government policies or programs.

In some sectors, such as agriculture, the collective decisions of multiple individuals are already strongly influenced by their understanding of current climate, and strong adaptive capacity has been demonstrated in the past (Easterling, 1996). Many assessments show that, as long as the impacts from climate change are not severe, the overall impact of climate change could be positive on agriculture, given the ability to change cropping patterns; to develop new varieties of crops and technologies for sowing, irrigating, cultivating, and harvesting; to use new sources of information; and the potentials for carbon fertilization of crops from higher concentrations of carbon dioxide (CO_2) in the atmosphere (Reilly et al., 2001). Nevertheless, some adaptation options in agriculture might be more difficult to implement due to the complex structure of the agricultural economy and international markets and increases in the rate of change.

There are many other private-sector opportunities to engage in adaptive behavior, including private consulting firms who specialize in assisting decision makers within various sectors in using climate information. Some consultants have already developed both the knowledge and the tools to perform detailed numerical analysis in support of adaptation planning for local, state, or tribal governments—for instance, in evaluating responses of water resources, energy, transportation, and agriculture to climate change. Citizens are already engaging through individual participation in conservation programs, by serving on public advisory committees, neighborhood groups and other volunteer activities, and getting involved in planning processes for adaptation.

The private sector brings significant capacity to assess risks, make decisions, and integrate new sources of information, including through the global marketplace (e.g., the United States Climate Action Partnership,[1] which has played a major role in integrating climate change concerns into the decision processes of major corporations; USCAP, 2007). Insurance companies, particularly those with global perspectives, are very focused on using the latest climate change information in assessing risks and developing their cost and benefit structures (e.g., Lloyd's of London, 2008), and financial and investment institutions also have reasons to include climate change risks in their investment strategies. Energy companies are focused on increasing the cost-effectiveness of renewable energy alternatives, on building an energy delivery system that is compatible with these sources, on optimizing their future economic strength in the context of emerging climate policies (*ACC: Limiting the Magnitude of Future Climate Change*; NRC, 2010c), and on assessing their options for ensuring that their delivery capacity is resilient in the face of impacts from extreme events and sea level rise (Chapter 3). All of these activities (which include emissions reduction as well as adaptation efforts) require adaptive capacity of multiple kinds, including the capacity to incorporate new information into complex decision processes. Where financial incentives exist for the private sector to promote adaptation, there may be very little need for government intervention.

Nonprofit organizations have played an important role in advancing adaptation planning at regional and local scales in the United States and abroad, and in many cases they have been leading by example. For example, the ICLEI U.S. Climate Resilient Communities Program, Royal Melbourne Institute of Technology Global Cities Institute: Global Climate Change Adaptation Program, and the World Bank have each sponsored efforts supporting local urban planning for climate change impacts (Pew Center on Global Climate Change, 2009). The Nature Conservancy has developed a major role in adaptation activities for habitat and species conservation in partnership with others. For example, they have partnered with NOAA's Coastal Services Center, Columbia University's Center for Climate Systems Research/Goddard Institute for Space Studies, Pace University's Land Use Law Center, and the Association of State Floodplain Managers in sponsoring coastal adaptation planning efforts in Long Island, New York (TNC, 2009). The Florida Coastal and Ocean Coalition, made up of numerous NGO members, developed a report for Florida officials outlining actions that can be taken to address climate impacts (FCOC, 2009). The Rockefeller Foundation has been focusing on adaptation efforts in developing countries, and the John D. and Catherine T. MacArthur Foundation has committed funds to conservation groups to protect biodiversity in

---

[1] *http://www.us-cap.org/*, accessed October 11, 2010.

ecologically rich "hot spots" around the world from climate-related impacts (as reported by Stutz, 2009). Clearly, these are only a few examples, and NGOs will continue to play an important role in advocacy and coordination of adaptation activities, along with encouraging broad societal support for climate change adaptation.

THE NEED FOR A COORDINATED NATIONAL APPROACH TO CLIMATE CHANGE ADAPTATION

As indicated in previous sections of this report, several local, state, and regional institutions in the United States have begun to engage in planning and implementing responses to climate change impacts, both governmental and nongovernmental. These examples are small in number relative to the overall number of jurisdictions, programs, sectors, and vulnerabilities; and there are even fewer examples of "comprehensive" adaptation plans that have attempted to take into account interactions across sectors and to prioritize resource allocations based on cross-sector vulnerability analyses. Comprehensive adaptation planning calls for the involvement of a large number of entities across all scales and types of decision makers. For example, several decision-making processes for which cities and counties are responsible (e.g., land use planning) are currently not represented at the national level and, unless they are addressed by a comprehensive national initiative, will be underserved in developing adaptive capacity (NRC, 2009a). There is currently no strategy, however, for a coordinated approach across scales. A "patchwork" of adaptation plans, actions, and capacities could result in inconsistent, conflicting, inefficient, or inequitable investments and responses and would be difficult to evaluate and monitor over time. In addition, climate impacts will often cross jurisdictional boundaries. No mechanism or policy approach currently exists to provide such coordination across boundaries between scales and different contexts for decision making.

While many decisions regarding adaptation will fall on local and state institutions, the federal government can uniquely assist, support, and coordinate America's choices related to adaptive risk management. Specifically, federal agencies can provide financial and technical resources, including information about climate change and climate change impacts. As demonstrated by the examples of Keene, New Hampshire, and Honolulu, Hawaii, and the statements of groups such as the Western Governors Association and U.S. Conference of Mayors, local and state organizations often lack sufficient resources when identifying, evaluating, and monitoring adaptation options. In addition, the federal government can coordinate efforts across its agencies, in conjunction with state and local entities. Of particular concern is to identify areas where competing or conflicting federal regulations inhibit adaptation. As shown by some of the international examples in this chapter, coordinated planning can create and improve opportunities for individuals, private entities, and boundary organizations to participate in and benefit from adaptive responses.

All of these roles for the federal government have recently been recommended by GAO (2009b), arising from its assessment of adaptation to climate change. The GAO report offers clear recommendations for the establishment of a "national adaptation plan," which would provide a framework for coordinating efforts among local, state, and federal entities, in addition to organizing and allocating financial and technical resources.

Two examples of coordination across levels of governance are offered in the next section. These demonstrate how an "intermediate" approach might be structured and implemented, combining elements of bottom-up decision making and financial, technical, and strategic support from higher levels of governance. Box 4.8 also discusses potential changes to the NFIP, which serves as an additional example.

Existing Models for Multijurisdictional Coordination

State and Local Hazard Mitigation Plans

There are several examples of existing legislation and programs that have provided assistance to state and local governments and other parties in adapting to and mitigating natural and manmade hazards. The first is FEMA's hazard mitigation program. This program provides assistance to states, territories, tribal governments, and local governments for long-term hazard reduction. The Robert T. Stafford Disaster Relief and Emergency Assistance Act of 1988 requires the states to develop hazard mitigation plans. These plans are developed in coordination with various state and federal agencies and with local governments. These plans provide the basis for grants made to the states under the Pre-Disaster Mitigation Grant Program and the postdisaster program, the Hazard Mitigation Grant Program. Under the FEMA Hazard Mitigation Program, FEMA as the lead federal agency provides guidance to the states and the states are responsible for coordinating and developing the plan.

The second example is the Superfund Amendments and Reauthorization Act (SARA). Title III of SARA is the Emergency Planning and Community Right-to-Know Act. The Environmental Protection Agency (EPA) coordinates the requirements of SARA Title III with federal, state, and local governments and private industry. SARA Title III requires states to establish a State Emergency Response Commission (SERC) to oversee the emergency planning requirements specified in the act. The SERC in turn requires the local governments to establish Local Emergency Planning Committees to coordinate compliance with the SARA Title III.

Federal Coastal Zone Management Act

Another model for intergovernmental coordination on a complex suite of planning and policy issues is found in the federal Coastal Zone Management Program, which has been administered by NOAA for more than 30 years. The Coastal Zone Management Act of 1972 (CZMA) established a unique partnership between the federal government and state and local programs to achieve both national and state priorities related to ocean and coastal issues. The act establishes national standards and program areas, with voluntary participation of the states. States were given a high degree of flexibility in developing their original plans and programs to meet the requirements of the act, to foster experimentation with unique approaches, and to account for diverse state and regional conditions. In return for participation in the program, states are awarded matching federal funds and the policy of "federal consistency" with approved state programs (i.e., federal activities must be consistent with approved state programs). All 35 of the coastal states and territories of the United States are currently participating in the program.

The CZMA, in fact, already authorizes limited funding for state and local programs for sea level rise planning activities. It also provides an interesting framework to consider with respect to adaptation planning—national policies and standards with strong focus on state-level planning and with provisions for federal consistency with approved state plans. For example, a state could plan (in consultation with local governments and stakeholders) for the future "armoring" of some coastlines to protect critical infrastructure and determine which shorelines should be allowed to transgress naturally over time. Once the state plan or policy was approved by NOAA, a federal project or facility could not be authorized if it was found to be inconsistent with that state plan.

CONCLUSIONS

Adaptations to impacts of climate change combine efforts by a wide range of U.S. institutions at different scales, from different sectors, and from different parts of the nation's institutional family: government, industry, and other nongovernmental institutions. Capacities for climate change adaptation are currently limited in most governmental and nongovernmental institutions in the United States at all scales and in all sectors.

A number of adaptation planning and implementation activities have been initiated by cities, regions, and states, providing opportunities to transfer lessons learned. At present, these emerging adaptation efforts in the United States are not well coordinated and could result in unintended consequences and inconsistent, inefficient investments and responses. Currently, there is no clear federal coordination or national strategy for climate adaptation. For a problem that crosses so many sectors and levels of government, that is so intricately woven into unique regional conditions and challenges, and that requires such significant public and private investments, integrated national coordination and clear strategies will be essential to (1) leverage limited resources; (2) avoid redundant or conflicting projects, mandates, guidelines, and assistance; (3) ensure responsible resource allocations over time and across scales and geographies; (4) improve understanding of changing conditions; and (5) encourage sharing of information, ideas, and lessons learned.

As a result, there is a clear need for increased federal engagement in climate adaptation efforts. The federal government has key responsibilities in addressing transboundary and interjurisdictional issues, providing scientific and technical support, advancing interstate commerce and national security, and protecting public infrastructure and lands.

> **Conclusion:** In keeping with recommendations of the U.S. Conference of Mayors, National Governors Association, and the findings of this chapter, the panel concludes that there is a need for the federal government to provide leadership by developing and pursuing a collaborative and inclusive national climate adaptation strategy.

> **Conclusion:** The impacts of climate variability and climate change are or will be felt within regions, communities, and sectors and are fundamentally place-based. It is not possible for the federal government to implement appropriate and cost-effective adaptation strategies without significant engagement of regional institutions, states, cities, tribes, and sectors. There is a need to significantly increase regional, state, and local capacities for adaptation planning and to make careful decisions regarding investments across sectors, impacts, and scales (see, e.g., Alaska case study in Chapter 3). Currently, capacities are limited in many areas, and significant funding and technical assistance will be required to build adaptive capacities at appropriate scales in the United States.

> **Conclusion:** A national adaptation strategy is needed to facilitate interstate and international cooperation with regards to adaptation planning, along with collaboration across lines between government and other key parties, and should

clearly articulate national interests and goals in climate change adaptation. Such a strategy should include effective institutional arrangements that consider the potential value of federal incentives (funding, technical assistance, intergovernmental consistency), standards, and requirements.

Conclusion: The national strategy would benefit from a "bottom-up" approach that builds on and supports existing efforts and experiences at the state and local levels and efforts of partners in the private sector and other NGOs. The strategy should be action- and results-oriented and should measure progress in terms of improving the nation's adaptive capacity, improving quality of life, and building economic advantages by finding solutions to high-priority climate change impacts and reducing risks and vulnerabilities.

Conclusion: The magnitude and complexity of the adaptation problem require forging new relationships among the public and private sectors, academia, interest groups, government agencies at all levels, and private citizens. In some cases, it may be most appropriate to develop adaptation plans that are sector-based, such as within the energy industry. In other cases, regional plans or programs may be more effective. The roles and responsibilities of decision makers at multiple scales will need to be defined and then refined over time.

Conclusion: A national strategy, implemented through a national adaptation program, is also needed to coordinate among federal programs, decision making, planning, and regulations and to "mainstream" considerations of climate change adaptation. Examples of programs where climate adaptation components, including financial and technical assistance, could be incorporated include the "Farm Bill" (and agricultural policies more generally), the NFIP, agency and program authorization bills, the National Environmental Policy Act, the CZMA, and the Endangered Species Act. In some cases, successful adaptation cannot be accomplished solely by "mainstreaming" climate change into existing programs; conflicts and constraints arising from federal mandates will require a reexamination of goals and requirements. A number of federal tax incentives or subsidies should also be reexamined in light of climate impact projections. The federal government should reexamine disaster relief, flood insurance, agricultural subsidies, and other influences to ensure that existing programs and policies do not result in further development of hazard-prone areas or maladaptive practices.

Conclusion: There is a need for more support for proactive strategies and planning processes that consider multiple perspectives and competing interests. Adaptation plans need to provide a flexible framework for setting priorities and coordinating implementation, including regional partnerships, and have to ensure strong public participation, as well as nongovernmental and private-sector stakeholder engagement in planning and implementation (see Chapter 4).

Conclusion: Public education and extension components are a critical part of a national program because effective adaptation measures will require the participation and support of individual citizens and a variety of sectors (*ACC: Informing an Effective Response to Climate Change*; NRC, 2010a).

Conclusion: Finally, a national adaptation program itself needs to be adaptive and continually strive to increase its own effectiveness. An ongoing assessment of progress (in terms of both outcomes and process) involves promoting change that is informed by ongoing information collection and dissemination, as opposed to a rigid response intended to be permanent. Other critical features of adaptive management include learning from past and emerging experiences, recognizing the complexity and the interrelated nature of sectoral interests such as water, agriculture, and energy and understanding the relationships between adaptation activities and the need to limit GHG emissions. Over time, there will be a need to adapt to our own adaptations (and maladaptations) as well as to our efforts to limit the magnitude of climate change.

CHAPTER SIX

Rationale and Mechanisms for Global Engagement in Climate Change Adaptation

America's climate choices regarding limiting greenhouse gas (GHG) emissions and adapting to climate change impacts will be implemented in a global context. At a fundamental level, the decisions made by individual governments are linked to impacts in other countries through the climate system, the global economy, and in many other ways. Each nation's climate change responses can impact systems that cross international borders, influencing everything from water flow in major river basins to population migrations and food supplies. Moreover, climate change impacts and adaptation actions—or the lack of thereof—in every country will affect competition in global markets for climate-sensitive products and services (e.g., forest products and tourism; Denman et al., 2007).

As international climate change negotiations under the United Nations Framework Convention on Climate Change (UNFCCC) draw attention to the need for adaptation as well as limiting GHG emissions, the United States, as the world's historic leading emitter of GHGs, will have strategic choices to make about how to engage and respond. As discussed in detail in the companion report *ACC: Limiting the Magnitude of Future Climate Change* (NRC, 2010c), the high-income countries have been the leading contributors to cumulative GHG emissions. However, emissions in the emerging economies (e.g., Brazil, China, and India) are projected to grow much more rapidly than those in developed countries. In fact, current projections indicate that the low- and middle-income countries will account for the bulk of cumulative global GHG emissions in the future (NRC, 2010c).

Therefore, choosing to engage in international dialogues and actions about climate change adaptation could have several benefits for the United States. First, it addresses questions of global equity with regard to developing countries bearing the consequences of climate change resulting from emission from the developed countries. Second, it is an opportunity for the United States to provide assistance for international humanitarian concerns as part of existing development goals. Third, international engagement can address national security issues that could arise from climate change. Fourth, coordination among countries could improve the effectiveness of adaptation efforts by reducing redundant activities or those that work at cross purposes and by facilitating an exchange of lessons learned. Fifth, international engagement offers the United States opportunities to learn from the adaptation experiences of others. And sixth, international engagement offers opportunities for U.S. adaptation technologies, systems, and services to find expanded markets globally.

This chapter discusses the international context for adaptation and concludes by highlighting the benefits of integrating climate change adaptation objectives into a range of foreign policy, development assistance, and capacity-building efforts. Such a climate policy can improve the United States' ability to influence a broader range of outcomes, including economic and national security considerations. Overall, the chapter highlights the importance of building solutions and making decisions on adaptation options within a broad international context (Bales and Duke, 2008; Bang et al., 2007; World Bank, 2010).

CLIMATE CHANGE IMPACTS IN AN INTERNATIONAL CONTEXT

Climate change is already affecting resource availability globally, and future impacts could lead to dramatic changes in economic and environmental conditions, creating both humanitarian and national security concerns (IPCC, 2007a; Khagram and Ali, 2006; World Bank, 2010). For example, projected increases in the frequency and intensity of extreme weather events could lead to increased vulnerability across the globe; these and other climate-related threats to sustainable development in some countries in Africa and Asia may create an increased need for humanitarian assistance (World Bank, 2010). In countries with unstable governments, climate change impacts can act as stress multipliers that have the potential to contribute to geopolitical instabilities (CNA, 2007). Particular concerns might include regional water scarcity and food shortages (Cooley et al., 2009; Schmidhuber and Tubiello, 2007), severe storms, sealevel rise in densely populated low-lying areas, and human health impacts of climate change (World Bank, 2010). In addition, the potential migration of populations that may be displaced by climate change (e.g., by sealevel rise or persistent drought) could exhaust resources available where resettlement is established (World Bank, 2010). Should any of these destabilizing events occur across borders, the resulting resource competition could possibly lead to international conflict.[1]

Specific examples of complex international issues related to climate change impacts that could potentially be addressed, at least in part, by adaptation include:

---

[1] In this report, "conflict" is used to refer to a subject of dispute between nations. It does not refer to military confrontations or war. "Violent conflict" is used to indicate military or armed confrontation.

- Food security in Asia and sub-Saharan Africa is being threatened by population growth, reduced soil fertility, changing dietary preferences, and other factors (World Bank, 2010). Additional increases in temperature and changes in precipitation could push some fragile regions over thresholds, resulting in further decreases in crop yields (Battisti and Naylor, 2009). Without adjustments to crop tolerances and yields, the risk of hunger could increase, and food scarcity could spur a large-scale human migration that could affect the stability of nations and the region. Adaptations to farm management practices and the development of crops resilient to change climate conditions along with improvements in crop yields will be needed to reduce food insecurity.

- Access to sufficient quantities of safe water for human, animal, and agricultural use in a changing climate is a major concern in many areas of the world. Increased precipitation is anticipated in some parts of the globe, but, even in places that get more rain rather than less, climate projections indicate that more of that precipitation is likely to come in a smaller number of intense rainfall events (Kundzewicz et al., 2007), posing risks of both flooding and seasonal or interannual drought. Although there is some debate about the degree to which water has been a cause of conflict in the past (see, e.g., Barnaby, 2009), increasing demand for water has the potential to fuel future conflict in arid regions (Cooley et al., 2009). Many countries are interdependent for their water supply because rivers cross borders. Conflict could arise among countries where increased variability in the water supply from climate change outpaces the ability of relevant institutions to adjust (Wolf et al., 2003). More international cooperation, possibly including new venues for collaborative discussions and water management, will be needed to avert conflict (Cooley et al., 2009; Wolf et al., 2003; World Bank, 2010).

- Climate change-related increases in the health burdens of malnutrition and diarrheal and other infectious diseases could lengthen the time required to achieve development goals by increasing resources needed to treat and control these health impacts, reducing worker productivity, and impairing childhood development. Improving health protection programs (e.g., malaria surveillance and control, increased attention to maternal and child health, and reduced risk of malnutrition) would increase the capacity of countries to avoid, prepare for, and cope with any changes in disease burdens (Confalonieri et al., 2007; Patz et al., 1996).

- Changes in the Arctic are affecting livelihoods and traditional ways of life and changing ecosystem diversity. Shrinking of Arctic sea ice is creating dramatic implications for access to resources, coastal erosion, the fate of arctic mammals, and the productivity of fisheries in the region (Post et al., 2009). As marine access to the ice-free Arctic Ocean increases, sovereignty, security, and safety issues will become more common (ACIA, 2005). To adjust to the conditions, adaptation will include the relocation of some communities. International dialogue and coordination regarding newly open water will also need to occur.

- The consequences of sea level rise for Pacific Island nations and Asia have the potential to displace populations across the globe. Handling such migration may require new international policies and approaches to facilitate migration and refugee resettlement (Gilbert, 2009).

RATIONALE FOR U.S. ENGAGEMENT IN ADAPTING TO CLIMATE CHANGE AT THE GLOBAL SCALE

The rationale for U.S. engagement in international adaptation efforts stems in part from equity considerations, associated with the fact that the United States has played a major role in the buildup of GHGs in the atmosphere to date. In fact, the UNFCCC Conference of the Parties (COP), including the United States, agreed to the Bali Action Plan in 2007, which formalizes a process for nations to join in an effort to support adaptation to climate change.

Global Equity Considerations

The UNFCCC is a global environmental treaty with nearly universal membership (192 parties) and includes nations that are the main contributors of emissions as well as those that will need to adapt to the impacts of climate change. Due to the greater socioeconomic vulnerabilities in developing nations, the poorest nations who have contributed the fewest GHG emissions will likely face the greatest consequences (IPCC, 2007a; World Bank, 2010). Consequently, during the COP in Copenhagen, the most important objective of international climate policy negotiations for many of these developing nations was obtaining a commitment from the developed nations on both GHG emissions reductions and substantial adaptation funding (SEI, 2009).

Although developed countries have already committed (Article 4.4) to assist developing countries in meeting adaptation costs, the available funds fall short of the estimated amounts of funds needed (Klein and Möhner, 2009). In addition, there are concerns as to whether the current system of funding adaptation is adequate, even if the funds were fully funded (Klein and Möhner, 2009; SEI, 2009). There are currently several funds established under the UNFCCC: the Global Environmental Facility Trust Fund, the Least Developed Countries Fund (LDCF), and the Special Climate Change Fund. There is also an Adaptation Fund under the Kyoto Protocol, which has never become operational. Adaptation financing is a major hurdle to making progress on an international agreement under the UNFCCC, and proposals for overcoming these obstacles have been offered (e.g., SEI, 2009). Although the rationale for engagement is apparent, the mechanisms by which to engage are diverse, as discussed below.

Advancing Development Goals

The impacts cited in the previous section are representative of the kinds of humanitarian crises that climate change is likely to exacerbate. The United States has the opportunity to assist with adaptation efforts that improve the capacity of these countries to adapt and advance general development goals. Examples of activities that the United States could engage in to advance these objectives include providing more aid in response to extreme events and other climate-related impacts (O'Brien et al., 2006; Schipper and Pelling, 2006), participating in the planning and management of transborder resources and human migrations (de Wit and Stankiewicz, 2006; Niasse, 2005; World Bank, 2010), increased monitoring of and support for food security (Schmidhuber and Tubiello, 2007; World Bank, 2010), and developing and transferring renewable energy technologies that reduce GHG emissions (Brewer, 2008; de Coninck et al., 2008).

U.S. National Security

National security implications of climate change for the United States have received increased attention recently (e.g., Blair, 2009; Broder, 2009; Busby, 2007; CNA, 2007). The concerns are twofold: First, as noted earlier, climate change will potentially have adverse and sustained impacts on resource availability and vulnerability in some regions and might disproportionally affect developing countries and the poor (Schmidhuber and Tubiello, 2007). Because of these adverse impacts, a recent report by the intelligence community concluded that the decreased availability of resources, such as water and food, represents a stress multiplier (CNA, 2007). Governments may face increasing difficulty sustaining their populations and maintaining political stability as they cope with resource stresses induced by climate change (CNA, 2007). Conflict over resources made scarce (or newly accessible) by climate change is a possibility. Climate change could affect the stability of nations and lead to conflicts that the United States may need to engage in. For example, transboundary water issues can result in increased tension; however, most often agreements are found (Wolf, 2007; see also discussion below). In fact, a gradual decrease in the resource after agreements are found is more likely than violent conflict. This gradual decrease in water quality or supply can cause internal instability and affect human welfare (Wolf, 2007). Diplomacy, intelligence, military force, and economic aid are components of an integrated and coherent strategy for addressing climate change adaptation in the context of national security (CNA, 2007). Future socioeconomic conditions in developing countries, however, are quite uncertain. Should standards of living rise, as they will in some countries, those nations' adaptive capacity is likely to increase (Tol et al., 2004).

Second, the effects that climate change will have on the operational environment represent a concern for national security, though the extent and timing of such impacts remain uncertain. Rising sea level, increases in extreme climate-related events, long-term changes in patterns of precipitation and temperature, and changes in access to and availability of resources are all factors that are relevant to the operational environment of the U.S. military at home and abroad. As an example, military installations along the coast are likely to be affected by sea level rise and may need enhanced measures to cope with more intense coastal storms. Military forces may also be called upon more frequently to respond to extreme events such as hurricanes, wildfires, heat waves, floods, winter storms, and drought. U.S. military forces already possess unique and unequalled capabilities to deploy and support such activities on a global basis and will continue to do so.

 For these reasons and others, the national security community has been increasingly concerned with climate change, which has been recognized as a threat to national security by several independent reports (e.g., Busby, 2007; CNA, 2007). The Navy is addressing these concerns through a high-level task force on climate change, and there is evidence that this subject will be addressed in considerably more detail in forthcoming documents such as the Department of Defense Quadrennial Defense Review (DOD, 2006), the National Security Strategy, the National Military Strategy, and the State Department's Quadrennial Diplomacy and Development Review (DOS, 2009). In addition, the Senate has held hearings on climate change impacts as they relate to the U.S. military and national security (Blair, 2009; Warner, 2009). Recognizing the security implications of a changing environment, the White House issued a national security presidential directive in January 2009 updating its policy related to homeland security and defense and the effects of climate change and human activity in the Arctic region (Presidential Directive, 2009). Where climate change poses challenges to U.S. national security, adaptation alternatives may provide strategic opportunities (Busby, 2007).

Transboundary Adaptation Challenges

As discussed previously in this report, adaptation activities have the potential to be redundant or to work at cross purposes if they are not coordinated across sectors, actors, scale, and time frames (Chapters 3-5). Adjustments made by one country can have unfavorable consequences for others. To examine the need for international coordination on adaptation, transboundary river basins serve as an example of the kinds of cooperation and dialogue that will be required to coordinate climate change adaptation. International river basins are a good proxy because they occur all over the world, the actors they link vary in number and in economic development, and there is an existing body of literature devoted to the sharing of resources over time in these locations.

There are 263 transboundary waterways in the world today, the watersheds of which include 40 percent of the world's population, 47 percent of the Earth's land, and 60 percent of the freshwater resources (Wolf, 2007; Wolf et al., 1999), including several rivers that cross U.S. boundaries. An example is the Colorado River, which flows through seven states and crosses the U.S.-Mexico border. Water issues in the Colorado River basin are particularly complex and have generated significant tension, both among U.S. states and between the United States and Mexico (Figure 6.1). Recent developments have showcased the benefits of both official diplomacy and the roles of nongovernmental partners, as well as more innovative approaches to problem solving and capacity building (Cooley et al., 2009).

The Colorado River basin is currently experiencing not only the worst drought in a century of record keeping but one of the worst droughts in more than 500 years (based on tree-ring data and other paleoclimate studies; Timilsena et al., 2007). Water levels in its two largest reservoirs, Lakes Powell and Mead, have plummeted from nearly full in 1999 to half empty due to low levels of runoff. Recent modeling studies of climate change and projected population growth in the basin indicate that the current drought conditions can be understood as a preview of the future (Milly et al., 2008; Seager et al., 2007). Climate change impacts are expected to include increased ambient temperatures, evaporation, and evapotranspiration rates; reduced and altered timing of precipitation and runoff; and increased energy demands. These impacts will add to water supply challenges associated with ongoing development in the Upper Basin (Colorado, Wyoming, Utah, and New Mexico) and rapid population growth throughout the Lower Basin (Arizona, California, and Nevada) and Mexico. Shortages represent a particular concern for Mexico because that nation lacks on- or off-stream storage and depends directly on U.S. deliveries to meet water supply needs for municipal, agricultural, and environmental uses. Population growth in border communities and along the Pacific Coast may result in critical water supply shortages for Tijuana,

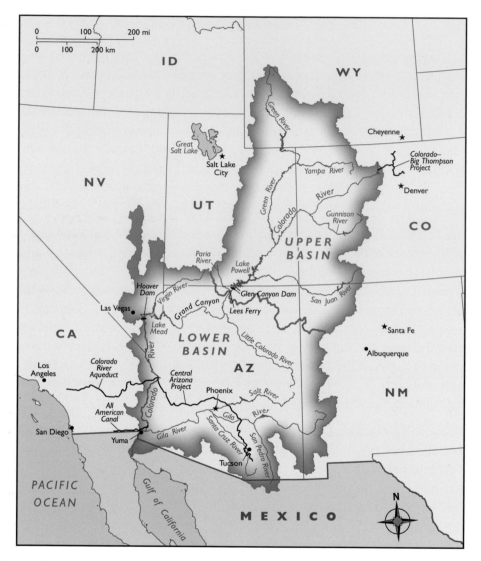

FIGURE 6.1 Map of the Colorado River basin. SOURCE: International Mapping Associates.

Ensenada, Mexicali, and other communities, which in some cases have already outstripped existing water supplies. Groundwater use and the continued availability of groundwater recharge remains a critical concern to agricultural users in the Mexicali Valley.

The United States and Mexico have long maintained formal international institutions for the management of the Colorado River. The U.S. International Boundary and Water Commission (IBWC) and its Mexican counterpart, the Comision Internaciónal de

Límites y Aguas (CILA), have existed for over 120 years to apply boundary water treaties between the two countries and to address differences that may result from the application of the treaties.[2] The fact that more than 300 clarifications and amendments have been made to the 1944 treaty that governs water delivery from the Colorado River to Mexico shows that IBWC and CILA have established a precedent of adapting to changing conditions in the basin (Cooley et al., 2009; Tarlock, 2000). Nevertheless, climate change is not specifically referred to in any of the amendments, nor are adaptations to water scarcity in both countries that might reduce stresses and buy time to develop effective cooperative management approaches for the longer term. These institutions have historically provided a forum for dialogue and cooperation, but they may still prove insufficiently flexible to address rapidly changing conditions related to climate (Tarlock, 2000).

Learning from Others

Many foreign governments have moved beyond a focus on limiting GHG emissions and are strongly engaged in climate change adaptation activities, participating in programs that range from funding and facilitating a coordinated adaptation research program in Australia (see Chapter 5) to supporting strong public engagement and capacity-building activities and efforts to bridge the science-policy interface (such as the United Kingdom Climate Impacts Programme; see Chapter 5, Box 5.5). Efforts in Germany provide a good example of an interdisciplinary, multisectoral approach to planning that focuses on the economic opportunities associated with adapting to climate change (see Box 6.1). Remarkably, some of the least developed countries have strongly supported implementation of adaptation efforts and offer opportunities for learning (see Box 6.2).

OPPORTUNITIES FOR U.S. ENGAGEMENT IN GLOBAL ADAPTATION ACTIVITIES

The United States has a wide variety of possible choices for incorporating climate change concerns into current international activities and considering appropriate U.S. roles in supporting global adaptation efforts.

---

[2] *http://www.ibwc.state.gov/home.html.*

BOX 6.1
Germany

In January 2009, the German Ministry of Education and Research announced its support for a consortium of universities, the regional Chamber of Commerce, and several stakeholder groups to develop the Northwest 2050 plan over the next 5 years (BMBF, 2010). This plan will include detailed analyses and comprehensive planning to prepare Germany's coastal northwest for the direct impacts of climate change on the environment, businesses, and society as well as the indirect impacts that may come from changes in markets, technology, migration, and other social and economic drivers. This adaptation planning focuses on three sectors: energy, food, and shipping.

Energy industry. The energy industry in the northwest region is intimately tied to national and international energy markets. While the region is projected to continue supplying energy from traditional sources, the region's coastline and landscape also offer ideal conditions for the development of wind and biomass energy sources. The energy industry is likely to be affected by climate change because of changes in availability of cooling water and more frequent storms that disrupt service and raise maintenance costs. Special attention will be given to the interplay of changes in energy technology, demand, and climate in order to inform investment decisions that improve the region's resilience.

Food-provision industry. This wide-reaching cluster of industries ranges from fisheries and agriculture to the supermarket. It is projected to undergo major adjustments due to climate change impacts. Agriculture will likely be affected as increased temperature and altered precipitation patterns change the crops that are planted and their yields. The forestry sector is projected to see a shift in ecologically

Opportunities for "Mainstreaming" Climate Change Adaptation Activities into Existing U.S. Programs

U.S. decisions about climate change adaptations at both national and international scales will take place in the context of existing treaties, trade relationships, resource extraction and processing networks, international development assistance, and global markets. This fact creates both constraints and opportunities: constraints because of the complexity of these interactions, and opportunities because many existing international engagement mechanisms can be used to achieve adaptation objectives. While it is beyond the scope of this report to provide a comprehensive review of or comment on the full range of international activities, climate change adaptation in the context of international development assistance is an example.

Climate change, if left unmanaged, has the potential to reverse development progress (World Bank, 2010). Agriculture and trade policies are linked to national and international food security. The U.S. Agency for International Development (USAID), which has a long history in international agricultural development, has engaged in a number of

supported forest types. Fisheries may be impacted by the projected loss of biodiversity. The supply chains for input into the industry and its outputs to markets may be disrupted by more frequent extreme weather events. Special attention will be given to issues surrounding supply chain management and public health under different socioeconomic and climate scenarios.

Shipping and trade industry. The region's ports are among its key economic engines—transporting goods internationally and supporting a large logistics industry. The international dimension of the shipping and trade industries and the manufacturing industries in the region require analysis of interrelated impacts that may be difficult to predict or adapt to. For example, infrastructure may be at greater risk from rising water levels and more frequent storms. Insuring and supporting the industry may become much more expensive. Markets for goods and services will be affected, leading to regional shifts and changes in the volume of exchanges, thus affecting the local economy. Special attention will be given to social and economic ripple effects associated with changes in the viability of shipping and trade in the region, as well as the fiscal impacts on the local economy that may result from (re)investment in infrastructure and from changes in competitiveness and profitability of the industry.

Detailed analysis and assessment of each of these three industries will be guided by a set of overarching questions involving expected trends and scenarios, potential opportunities, desired outcomes, challenges, and future strategies that need to be implemented. The project will combine basic research into the application of complex systems theory (definition and assessment of vulnerabilities and resilience) and nonlinear dynamic modeling (exploration of macrobehaviors and self-organization under alternate boundary constraints) with extensive dialogue with stakeholders in the local and regional investment and policy decision-making communities.

activities to incorporate climate change adaptation into its mission and goals. Primarily, it attempts to provide stakeholders with Earth observations and climate information to allow for early warning (e.g., the Famine Early Warning System Network) or assist with disaster response. USAID's climate change adaptation program also assesses development projects to ensure they function under future climate change. Because developing nations are more vulnerable to climate change and have less adaptive capacity, meeting goals of poverty reduction and sustainable development will become increasingly challenging with climate change (World Bank, 2010). Providing enough clean water and food with projected population growth and climate change will require a coordinated and global effort with leadership from the developed world (World Bank, 2010).

The Office of U.S. Foreign Disaster Assistance (OFDA)—as part of USAID—is responsible for facilitating and coordinating U.S. emergency assistance overseas and funds activities to reduce the impact of recurrent natural hazards and provides training to build local capacity for emergency management and response. OFDA responds to all types of natural disasters, including earthquakes, volcanic eruptions, cyclones, floods,

BOX 6.2
The Least-Developed Countries Fund

The severe impacts of climate change will disproportionately threaten the most vulnerable parts of the world, including the poorest developing countries, countries in sub-Saharan Africa, and island nations. Roughly 100 countries with a population of almost a billion people fall in these categories. The Least-Developed Countries Fund, managed by the Global Environment Facility, provides financial and technical assistance to these countries to develop National Adaptation Plans for Action (NAPAs). Forty such plans have been completed so far, focusing mostly on agriculture, water, health, and capacity building. NAPAs use a common template and a "least developed countries expert group" that provides guidelines on impacts, vulnerabilities, and adaptation options. The NAPAs have received high marks for being country-driven in their development. Many NAPAs propose what look like standard development projects. However, to be considered as adaptation projects they needed to incorporate climate change into the design and demonstrate connections to the scientific evidence about impacts.

Bangladesh's adaptation plans can illustrate some early successes resulting from the NAPA process. Despite their resource limitations, some less developed countries are in the forefront on adaptation. Bangladesh (see figure below), which is especially vulnerable to sea level rise and tropical cyclone activity, has committed the equivalent of tens of millions of U.S. dollars toward development of a national adaptation strategy and is now building a program involving all government ministers (S. Huq, personal communication, May 4, 2009).

In addition to the focus on implementation of adaptations to sea level rise and tropical cyclones, Bangladesh is exploring new modes of engagement to support adaptation policies and funding. For example, 140 international representatives attended a recent conference in Bangladesh to discuss community-based climate change actions, ways to connect researchers to practitioners, and development of pilot projects and assessments (S. Huq, personal communication, May 4, 2009). A concept explored in this discussion was ways to engage young researchers and local knowledge in developing countries in building adaptive capacity. This could result in the creation of an international network of trained "capacity builders" who can support practical ground-up adaptation efforts without the need for major international negotiations and funding.

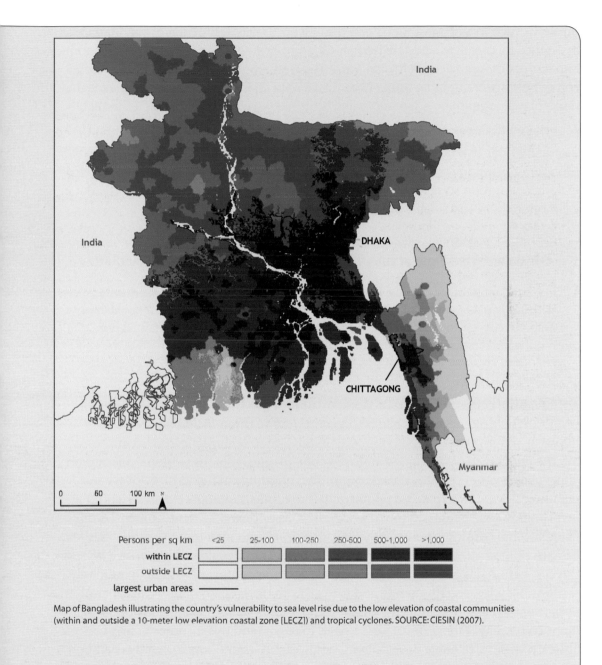

Map of Bangladesh illustrating the country's vulnerability to sea level rise due to the low elevation of coastal communities (within and outside a 10-meter low elevation coastal zone [LECZ]) and tropical cyclones. SOURCE: CIESIN (2007).

droughts, fires, insect infestations, and disease outbreaks. It provides humanitarian assistance to save lives, alleviate human suffering, and reduce the social and economic impacts of emergencies worldwide. Because weather extremes are projected to become more severe or more frequent as the climate changes, OFDA provides a mechanism for mainstreaming adaptation efforts by making additional resources available.

In addition to development programs, a wide range of international programs and a large number of bilateral or multilateral U.S. free trade agreements are available as tools for promoting environmental standards. Such trade agreements could therefore include adaptation standards as part of their environmental standards in particular in promoting more sustainable agricultural practices.

Considering the Role of the United States in Global Adaptation Activities

In fact, adaptation may involve significant transformations rather than incremental changes, some of which will be painful to those in societies reluctant to, or not able to, embrace change. International action and funding may be required to assist in promoting resilience, not only to finance adaptation projects, but also to facilitate the exchange of knowledge and practices that embrace a resilience approach to adaptation. (Adger et al., 2009)

As previously discussed, several adaptation funds have been established under the UNFCCC. The United States is eligible to contribute to three, including the LDCF (Box 6.2). The first contribution from the United States occurred in 2009, demonstrating the reluctance of the United States to engage directly in these global adaptation efforts, while other countries have contributed hundreds of millions of dollars and continue to look for expanded opportunities to fund adaptation. Documenting that investments have actually gone to "adaptation" as opposed to "development" is difficult in some cases; it is particularly difficult to define an increment that qualifies for climate change adaptation funding beyond benefits to development per se. Although this distinction has been considered necessary in the context of assessing national responsibility for climate change and the call from developing nations for reparations as part of the climate change negotiations, adaptation is deeply linked with development co-benefits. Therefore, there is a need to reexamine this dichotomy, align adaptation with sustainable development goals, and consider how to avoid separating adaptation from other development activities.

Other options for U.S. engagement exist through a range of United Nations bodies, including the United Nations Food and Agricultural Organization, the United Nations Development Program (UNDP), and the United Nations Environmental Program (UNEP). For example, the UNDP has developed Adaptation Policy Frameworks with Global Environment Fund (GEF) support; and the UNEP's World Meteorological Organization and the Intergovernmental Panel on Climate Change have jointly conducted Assessments of Impacts and Adaptations to Climate Change (AIACC) in Africa, Asia, and Latin America, also with funding from the GEF. "AIACC aims to enhance the scientific capacity of developing countries to assess climate change vulnerabilities and adaptations, and generate and communicate information useful for adaptation planning and action" (AIACC, 2010). In addition to funding from GEF, AIACC has been directly supported by USAID, the U.S. Environmental Protection Agency, the Canadian International Development Agency, and the World Bank. In fact, the World Bank is an important mechanism through which the United States supports development and adaptation activities. The World Bank provides technical and financial resources to developing countries with the goal to reduce poverty. In addition to providing low-interest loans and grants, one of the World Bank's core functions is capacity building, through such components as the World Bank University, and it is an important contributor to climate change adaptation. The recently released report on "Development and Climate Change" (World Bank, 2010) exemplifies how it provides leadership and assistance on key topics. In addition, the World Bank Pilot Project on Climate Resilience (PPCR) contributed more than $500 million in less than a year to fund priority adaptation projects in ten least-developed countries.

The rationale for and importance of U.S. engagement in adaptation activities at the global scale have already been articulated. However, the U.S. government has yet to commit substantial funds to the adaptation funds of the UNFCCC. Negotiations in Copenhagen highlighted the need for a strong U.S. commitment to these funds in addition to its leadership and engagement in other climate change and international development programs. There are a number of considerations that need to be evaluated, including making determinations about what the primary goals of future engagement should be. It will be important to answer a number of questions about the priorities of international adaptation investments relative to investments within the United States and about the strategic implications of investing in particular adaptation projects and funds (see suggested questions for consideration in the "Conclusions" section of this chapter).

CONCLUSIONS

There are many reasons for the United States to engage in international climate change adaptation dialogues and actions. Some are benevolent, and some are self-interested; but in combination they make a compelling case for a significant U.S. role in international climate change adaptation.

The following critical questions need to be addressed in the context of developing a national adaptation program that recognizes the global context, reflects the lessons learned in other countries, and focuses on capacity building:

- To what degree should the United States address adaptation as a global issue rather than as an issue to be addressed by each country individually?
- How can adaptation programs appropriately consider the wide range of international issues, including national security, economic security, and sustainability of human and environmental systems?
- Which adaptation activities need to be coordinated with other countries to be effective? To what degree should adaptation activities be done collectively across boundaries?
- Where would it be most effective to focus on reducing the vulnerabilities that underlie potential climate change impacts? How can these vulnerabilities be addressed proactively so that impacts are delayed or prevented?
- What processes can be established to ensure that the United States benefits from lessons learned in adaptation in other countries? What types of global support systems could be most effective in reducing the impacts of climate change?
- Which adaptation activities are transnational in nature—for example, managing cross-border resources that are impacted by climate change?
- Are there adaptation options that other countries might pursue that would increase U.S. vulnerability? Are there actions that we might take that would increase vulnerability elsewhere, including creating national security, economic, equity, and competition concerns? Climate changes affect economic competitiveness differently depending on a variety of factors, including geography, resource base, and adaptive capacity.
- How can limiting of GHG emissions and adaptation activities be linked in a global context so that both are optimized? How can the United States work with other countries to prevent unintended consequences of these activities?
- How can the United States best support international adaptation to rapid change versus long-term gradual changes? It is clear that responses to rapid change in the near term are more likely to be in an emergency management context, whereas responses to gradual changes may be more appropriate for longer-term global conversations.
- What kind of institutional structure should support these conversations? There is a need for expanded engagement beyond emissions-limiting discussions,

including broad-scale international funding for adaptation, knowledge sharing, and technology. Should this be built into the United Nations system or another existing structure, or should a new institution be developed? Is the Least-Developed Countries Fund a good model?

- Historically, the world's economies were focused on resource extraction. Climate change affects primary productivity, the ability to produce and extract natural resources that are still the source of livelihood in vast areas of the world, and also national wealth. How can we predict such vulnerabilities and prepare for them?

- What principles should the United States follow in developing its emissions-limiting and adaptation strategies to ensure that international engagement is conducted in a consistent and constructive manner?

Adaptations that increase resilience in one nation might increase vulnerability in other nations; such interaction effects need to be evaluated as a part of U.S. engagements in international adaptation actions. For example, upstream nations may wish to build more reservoirs to capture declining water supplies in drier regions. Such action would likely reduce water supplies for downstream counties and could increase the potential for conflict (Cooley et al., 2009).

Conclusion: The United States would benefit from engaging in international adaptation activities as part of the UNFCCC. Through this venue it could address many of the above questions. U.S. participation in multinational institutional arrangements to coordinate adaptation activities across the globe could result in many valuable outcomes, including the development of a clearinghouse function to provide information about adaptation activities.

Conclusion: "Mainstreaming" adaptation considerations into a range of U.S. international activities that could abate exposure to vulnerabilities from climate change.

Conclusion: There is a need for new forums that allow for mutual exploration of techniques and technologies that support adaptation actions and for communication and trust building between the United States and other countries. Such collaborative forums could also be used as focal points for peer-to-peer conversations on science, geoengineering, and emissions-reduction strategies and policies. Other roles could include assessing the interrelationships within and between climate change policies, facilitating development of new adaptation technologies, collaborating on needs for observing systems and attribution methods, and facilitating ongoing research-to-action and implementation of adaptation activities.

CHAPTER SEVEN

Major Scientific and Technological Advances Needed to Promote Effective Adaptation to Climate Change

Amer-ica's climate choices with regard to adaptation are undermined by the fact that the nation has a limited base of adaptation knowledge, tools, and options related specifically to climate change for two reasons. First, in most cases, evidence of impacts of climate change is just beginning to emerge, so the effectiveness of adaptation actions cannot yet be evaluated. While options available to the nation for adapting to the impacts of climate change have in many cases been identified, the scientific understanding of the *effectiveness* of these options is lacking, given that climate change is likely to pose challenges beyond those that have been addressed in the past as adaptations to climate variability. Thus, the need for scientific and technological advances is pervasive across the field of climate change adaptation research.

Second, climate change adaptation research to date has not been a national priority (NRC, 2009a). In fact, adaptation has been such a low priority that "adaptation" does not appear in any current metrics for reporting climate change research (e.g., not in the budgets of the National Science Foundation and not in the budget summaries in the annual *Our Changing Planet* report of the U.S. Global Change Research Program [USGCRP] Climate Change Science Program). Recently, examination of the Climate Change Science Program has shown that investment in "human dimensions research," including but not mainly oriented toward adaptation, and nonresearch expenditures on decision support represent about 2 percent of the total climate change research effort (NRC, 2009c). Investment in adaptation research is only a fraction of that 2 percent.

Science and technology advances are therefore needed to respond to many questions now being asked by decision makers—questions related both to potentials for self-initiated adaptation by households and businesses and to likely needs for planned adaptation at all levels of our government and society. To realize the potential of adaptation, scientific and technological advances (including both near- and longer-term fundamental research) are needed to support adaptation analysis and assessment, to identify and develop adaptation options, and to strengthen adaptation management and implementation.

SCIENCE AND TECHNOLOGY ADVANCES TO SUPPORT ADAPTATION ANALYSIS AND ASSESSMENT

The first step in adapting is understanding a system's vulnerability to climate change impacts and assessing adaptation options to address these vulnerabilities. Advances in science and technology of several kinds are needed to support such analysis and assessments.

Improved Information

Science and technology advances to support long-term adaptive risk management regarding climate change impacts. Responding to climate change adaptively requires a continuing flow of information about impacts that are emerging and impacts that are being projected with a higher level of confidence (also see "Observing Systems"), and it requires a continuing flow of information about experiences with adaptation actions. Moreover, it requires effective mechanisms for ensuring that the information reaches adaptation decision makers in forms that are useful to them (*ACC: Informing an Effective Response to Climate Change*; NRC 2010a).

Tools to create place-based geospatial assessments of vulnerability that identify and assess especially vulnerable areas, sectors, and groups as well as appropriate adaptive responses (Chapters 2 and 3). For example, a redefinition of flood-event return periods is needed to improve the geospatial resolution of this hazard in the context of climate change and contribute to adaptation planning.

Improved information about climate change impacts under different assumptions about multiple driving forces and stressors (Chapter 2). For example, Hurricane Katrina and other recent experiences with natural hazards have demonstrated that vulnerabilities are shaped by social and economic conditions as well as by weather or climate phenomena, and the recent report *Global Climate Change Impacts in the United States* (USGCRP, 2009) shows that many long-term vulnerabilities are shaped by population size and distribution. Research is needed to match climate change impact projections with other driving forces of vulnerabilities, including technological and institutional change.

Improved Understanding of Impact Thresholds

Advancing understanding of thresholds or tipping points for climate change impacts, which in turn helps to determine the limits of adaptation (Chapter 2). To illustrate this challenge, one could envision the case where the projected rise in sea level along the U.S. Gulf Coast by 2050 (reflecting a combination of actual sea level rise and land subsidence; CCSP, 2008a) would be so large that sea walls and other adaptation efforts may not be sufficient to protect the region's traditional ways of life. Faced with either abrupt or gradual change, localities and sectors face the possibility that—at some level and/or rate of emerging climate change impacts—current human and environmental systems may become unsustainable. This phenomenon is poorly understood, but it is critically important for adaptation planning for relatively severe climate change scenarios. Understanding possible tipping points can help to inform adaptation choices to avoid reaching such points; it can help to design observational strategies and programs to monitor system changes to provide early warning of an impending threshold in time to consider adaptation options; and in some cases it can clarify limits of adaptation, perhaps pointing to the need for structural changes such as voluntary relocation inland.

Improved Knowledge of Behavioral Dimensions of Adaptation

Human behavior that affects prospects and avenues for adaptation. Except for autonomous adaptations by natural ecosystems, all adaptation actions depend on human behavior, and there is a critical need for research on determinants of adaptation that focus on this topic (Chapter 4). Scientific knowledge of human behavior as a factor in climate change adaptation is currently very limited (NRC, 2009b), but is a critical component to understanding how adaptation decision making might work and constitutes an important part of the knowledge base for adaptation planning and action at a variety of levels and sectors.

Institutional behavior that affects adoption and implementation of adaptation strategies, as well as monitoring of the effectiveness of these adaptation strategies. Because institutions shape climate change responses and because conditions for institutional innovation and change are fundamental to adaptation, this research area is considered a high priority (NRC, 1999, 2001a, 2005a, 2009a).

Improved Understanding of How Climate Change Adaptation Relates to Broader Sustainability Issues

How climate change adaptation relates to broader societal concerns about resilience and sustainability (Chapter 3) in the face of multiple threats, stresses, and opportunities. As indicated above, adaptation to climate change impacts is only one of many concerns for localities, sectors, and populations concerned about their futures. Other driving forces include changes in population trends, the global economy, technology, and institutions (IPCC, 2007a). It is critically important to identify strategies and actions that support adaptation to longer-term climate change and also provide benefits for resilience and sustainability in the near term.

Improved Understanding of Interdependencies among Climate Change Impacts and Implications of Adaptive Responses in One Sector to Other Sectors and Infrastructures, as well as to Economies and Societies More Broadly

Climate change impacts on one sector can affect other sectors as well. For example, increased water scarcity can affect energy production, agriculture, water quality, and urban growth. Reduced electricity production can affect water pumping and wastewater treatment. Even within a single sector, disruption of service in one segment can stress other segments as well. Effective adaptation to impacts of climate change will require a much more integrated approach to analyzing impacts and responses.

More fundamental research over a longer term is also needed to close a variety of gaps in the science of impact and vulnerability assessment. These gaps currently limit the ability to perform many kinds of adaptation assessments and option evaluations. Needs for fundamental science and technology advances include: improved information about climate change impacts under different assumptions about multiple driving forces and stressors, improved information about costs of impacts and both costs and benefits of adaptation options, and improved capacities to assess and represent uncertainties (Chapter 2).

SCIENCE AND TECHNOLOGY ADVANCES FOR ADAPTATION OPTION IDENTIFICATION AND DEVELOPMENT

In too many cases, decision makers and stakeholders concerned about climate change risks and interested in possible adaptations have difficulty finding information about their options, as well as the possible effects, costs and benefits, and limits of those options. Historically, most of the attention to adaptation research, as reported in the Intergovernmental Panel on Climate Change Third and Fourth Assessment Reports, has been focused on understanding differences in adaptation capacity (Chapter 5), that is, determinants of the inclination and ability of a party to adapt. Summaries of possible adaptation options in selected sectors include Wilbanks and Sathaye (2007), Bierbaum (2007), and a set of "Synthesis and Assessment Products" produced by the U.S. Climate Change Science Program, 2007–2009 (4.1 through 4.7), along with a workshop held at the National Research Council on October 25, 2007.

Adaptation Data and Decision Support

In the short term, a high priority for science and technology advances is to create an adaptation database that offers the best available information about adaptation alternatives, their characteristics, and examples of best practices. The database should contain information about costs of impacts and both costs and benefits of adaptation options, as well as measures of effectiveness as they become available. As part of monitoring the implementation of adaptation options, information needs to be gathered about the decision and regional context and the socioeconomic considerations to enable the transferability of "lessons learned" about the adaptation options. In time, such a database should be interactive, ideally offering users the opportunity to go beyond simply surveying existing answers and to ask questions about database entries and consult experts about issues not covered by those entries. For example, the lists of options in Chapter 3 can be viewed as a beginning of such a database; however, more careful analysis would be required before choosing among these options.

Sectoral Priorities for Science and Technology Advances in Support of Adaptation

Advances in science and technology that would significantly strengthen adaptation planning and implementation in selected sectors, drawn from Chapter 3 and the matrices, include the following (*ACC: Advancing the Science of Climate Change*; NRC, 2010b). The list was developed from the references listed above, along with the knowledge of sectoral experts on the panel (see Chapter 3):

Water

- Analyses of approaches for adapting water rights systems and practices to new climate conditions;

- Improved understanding of groundwater dynamics and recharge in the context of climate change;
- Technologies and practices for water-use efficiency improvement in multiple sectors;
- Lower-energy and renewable energy approaches to desalination of ocean water and brackish groundwater; and
- Assessments of the ecosystem and human health implications of reusing municipal wastewater.

Ecosystems

- Advances in estimating values for ecosystem services as a basis for assessing benefits and costs of adaptations, including possible losses of services valued by society;
- Integrated analyses of alternative approaches to promoting ecosystem stability under changing climate conditions—approaches that range from easing impacts by reducing other ecosystem stresses to enhancing natural or assisted species migrations (including invasive species issues in destination areas);
- Improvements in the science base for dynamic spatial ecosystem planning— for example, in oceans and ecosystems in areas facing stress from land use changes; and
- Improved maintenance of services from coastal or marine ecosystems, such as flood attenuation and water filtration.

Health

- Significant advances in the capacity to model health impacts of climate change, such as changes in geographic range of diseases and disease vectors;
- Advances in science and technology to reduce vector populations that would otherwise benefit in some regions from climate change;
- Contingency planning for responding to multiple concurrent health threats with limited public health care resources;
- Analysis of alternatives for improving the resilience of health care facilities and systems to major weather events; and
- Analysis of equity considerations—who actually bears the health-related costs of climate change and receives the benefits of adaptation actions.

Agriculture

- Analysis of implications of regional drying for the long-term availability of irrigation for agriculture;
- Advances in the understanding of climate change effects on pests and pathogens;
- Analysis of possible roles of microinsurance in agricultural risk management; and
- Cross-cutting analysis, such as impacts of heat on the productivity of agricultural workers, impacts of higher ozone concentration on crops, and relationship of changes in agricultural productivity to sustainability in developing countries.

Energy

- Improvements in the efficiency and affordability of space cooling technologies for buildings to assist adaptation to warming;
- New approaches for cooling thermal electric power plants that are significantly less water consumptive than most current practices;
- Analysis of electricity transmission and distribution systems to determine possible vulnerabilities to heat waves; and
- Implications of new climate change policies on energy choices, their effects, and adaptation alternatives.

Transportation

- Advances in developing materials for transportation systems that are less vulnerable to damage from temperature increases and water submergence;
- Improvements in the understanding of effects of climate change on regional and local hydrology in order to guide changes in infrastructure specifications (e.g., bridges and culverts); and
- How to design and operate transportation systems that function well in emergency response and evacuation.

Coastal Vulnerabilities

- Advances in the understanding of benefits, costs, and broader implications of alternative approaches to reduce vulnerabilities (e.g., protect with barriers, protect with stabilization and facility hardening, insure, or relocate);

- Improve the understanding of factors that influence decisions about coastal land use;
- Improve the understanding of the likelihood, causes, and implications of coastal eutrophication and hypoxia; and
- Examine options for sustaining coastal wetlands and ecologies with a rising sea level and more extreme storms.

International

- Development of technologies such as salt-tolerant crops and solar cooling that will assist climate change adaptation in lower-income countries;
- Enhancement of monitoring systems to detect sustainability stresses and to provide early warning of possible needs for adaptation in order to avoid or delay economic/environmental tipping points;
- Evaluation of alternative institutional mechanisms for supporting climate change adaptation and capacity building globally; and
- Identification of potential ancillary benefits of intelligence information generated for other purposes that could be used to inform adaptation strategies.

Cross-Cutting Science and Technology Needs

In addition, many urgent research needs cut across sectors. For example, each sector-specific action or plan needs to consider cross-sectoral interdependencies and interactions, where impacts and adaptation alternatives in one sector have implications for others as well. This should be explored, at least in part, in a place-based context such as one or more major urban areas, or one or more watersheds, where interactions can be traced in some detail.

Improved understanding about when to implement adaptation actions is urgently needed—for instance, in which cases should adaptation be started now to ensure longer-term resilience instead of waiting for impact uncertainties to be reduced? Timing issues are likely to differ among adaptation needs and options, but they include such considerations as whether adaptation now can be more participative and less expensive than emergency-based, reactive, and sudden problem solving; whether adaptation now is more likely to be placed in a broader sustainability context; and how important current uncertainties are in valuing investments in risk management.

Furthermore, research is required to evaluate options for encouraging voluntary relocation from high-vulnerability areas such as retreat from vulnerable coastlines. Alaska is already facing requirements to relocate vulnerable populations from affected coastlines (Chapter 3), and other areas in the United States are almost certain to face difficult choices over the next half-century as well—choices between expensive large-scale protection of current land use practices versus relocation of such land use to other areas. The most illustrative examples include Alaskan shoreline erosion and the threat of a disruptive rise in sea level along the Gulf Coast (CCSP, 2008a), mentioned above. Other areas subject to severe water shortages, wildfires, flooding, or sea level rise may face similar challenges in the future. A research literature is beginning to appear about retreating from especially vulnerable areas (Cutter et al., 2007; Kates, 2007), motivated by observed trends in impacts, along with rising costs of insurance and other factors. Improving the understanding of how to encourage this kind of voluntary risk-management assessment and decision making is essential in order to avoid potentials for socially disruptive and economically expensive problem solving in the mid- to long-term future.

In addition, new approaches to cost sharing for low-probability, high-consequence extreme events will be needed. Climate change impacts and vulnerabilities are likely to lead to large losses in buildings, infrastructure, economic activities, and ecosystem services (Chapter 2). While in theory some of these are preventable, in practice the most common experience—that is, the most common adaptation—will be the bearing or sharing of these losses. Research is urgently needed on the distribution of these losses, the current modes of sharing (disaster relief, insurance, and government reimbursement), and new methods of sharing in the form of comprehensive or specialized climate insurance, catastrophe trust funds, and the like. Objectives include determining the best balance between investments in local adaptations and investments in cost sharing in the event of low-probability, high-consequence events. This should involve (1) both avoiding inefficient redundancies in local self-insurance against low-probability contingencies and realizing potentials for local adaptations to reduce the cost of insurance, and (2) exploring the right balance between private- and public-sector approaches to cost sharing.

SCIENCE AND TECHNOLOGY ADVANCES FOR ADAPTATION MANAGEMENT AND IMPLEMENTATION

Climate change adaptation is not just a set of actions. It is an ongoing process of learning and adapting, both to emerging information about climate change impacts and to evolving experience with adaptation strategies and decisions. Science and technology advances are needed to improve that process in the following areas (Chapters 3 and 4).

Areas for Science and Technology Advances

Observing systems. Deploying and using systems are necessary to monitor climate change impacts and to provide a continuing flow of information about them to decision makers (see also "Improved Information," earlier in this chapter). Developing an adaptive approach to adaptation and reassessing risk management as additional information emerges about impacts and responses depend fundamentally on this kind of accurate and timely information.

Risk analysis and management approaches. The development of risk analysis and management approaches is needed to provide tools and guidelines for adaptation decision makers (Chapter 4). Examples include refining and testing approaches for addressing such issues as the treatment of risks, the timing of adaptation actions, and benefits and costs of alternative actions, and for understanding of the likely winners and losers associated with any particular adaptation strategies and actions.

Learning from emerging experience and documenting and disseminating best practices. In nearly every case, at nearly every scale, there will eventually be a need to assess the outcomes of adaptation strategies and actions, asking such questions as what they have accomplished and whether they can be considered successes. The current state of the art and science for such assessments is very limited and is complicated by such factors as the need to define a baseline for comparison and the fact that multiple factors will influence the observed outcomes of any particular adaptation action. To build the knowledge base that allows the development of "best practices," the various factors that influence the choice and the outcome of the adaptation option need to be measured, archived, and analyzed. Ideally, a standard set of variable is monitored in every case to enable analyzing the link between option and geographic context. This is a significant cross-cutting adaptation research need in a field where, in at least some cases, practice is likely to proceed in parallel with knowledge enhancement, including needs to improve knowledge of the costs of adaptation.

Best practices in adaptation management (for example, successes with mainstreaming adaptation in ongoing community and sectoral processes, including institutional structures that sustain attention to adaptation beyond the spans of attention of individual leaders, and successes in improving resilience to climate-related disasters) need to be identified, documented, and disseminated. Validating "best practice," of course, depends on monitoring the effectiveness of policies and practices as they are implemented (see "Improved Information" and "Observing Systems" in this chapter).

Monitoring the impacts of adaptation and mitigation actions; adapting to adaptation and mitigation actions. Adaptation does not stop with adapting to direct changes in climate alone; it also faces challenges in adapting to what people do in responding to climate change. That includes actions to limit greenhouse gas (GHG) emissions as well as adaptation actions (and potentially, geoengineering options). Because these second-order adaptations have received very little attention, integrating them into adaptation planning and actions will be a challenge. For example, emissions-reduction strategies that raise the cost of some forms of energy, or geoengineering strategies that could have secondary consequences, may call for advances in adaptation research to support assessment, option identification, and option implementation (Chapter 2). In addition, some actions aimed at adaptation may in fact represent "maladaptations." A frequently cited example is what has been called the "levee effect" (e.g., Kates et al., 2006), where short-term adaptive responses create a sense of security and lead to societal responses that increase the chances of catastrophic risk in the future when short-term adaption options become inadequate.

Strengthening the Science and Technology Base for Adaptation

Optimizing the nation's adaptation to impacts of climate change is likely to require more than advances in science and technology in subject areas that can be identified now. It will most likely require transformational changes in our science and technology base for adaptation, too. This means that a national science and technology enhancement strategy should include investments in more innovative, "farther-out" ideas as well familiar ideas. For example, most adaptations to climate change considered today are extensions of existing options for adapting to climate variability or extreme events, differing only in the scope of implementation, the frequency of application, and the intensity of effort. But climate change may well exceed the range of current climate variability and extreme events; thus, novel adaptations are very likely to be needed, especially in the event of tipping points and/or abrupt changes. Particularly in the case of such potential severe or unexpected consequences, prudent risk management suggests the need to consider contingency plans for high-impact, low-probability events in adaptation planning. However, much research is needed on how to develop such "worst-case" plans.

Finally, there is a need to assess options currently not feasible. A number of needed adaptations that should be considered in national strategies seem infeasible today because current opposition outweighs their possible longer-term value. For example, many recognize the long-term need for a retreat from the coast but are unwilling to pursue this option, given society's heavy investment in current coastal development. Similarly, the future climate system is unlikely to supply the growing water demands of the South and West, meaning that water rights surely need revision. Given the history and diversity of water rights and the major difficulties experienced in pursuing even modest revisions, however, few decision makers are willing to undertake it. How such unfeasible solutions might become feasible should be the focus of a specific research effort. A research program designed to elicit such innovations is desirable.

ALTERNATIVE APPROACHES FOR MEETING SCIENCE AND TECHNOLOGY NEEDS FOR CLIMATE CHANGE ADAPTATION

How can these high-priority needs for science and technology advancement best be met? It is not the place of this report to prescribe a specific mechanism and budget level, but the panel suggests some possible guidelines for developing effective research mechanisms.

Possible Guidelines

Forceful actions to show that the nation considers adaptation a high priority and wants to improve its science and technology base in order to achieve adaptation goals as effectively and efficiently as possible. The need for a higher level of science and technology effort in support of adaptation has been identified previously (GAO, 2009a,b; NRC, 2009a,b), could be addressed by increasing its visibility and emphasis in government priorities, and might require changes in organizational structures in federal agencies and increased level of funding.

Involving a wide range of science and technology users and stakeholders in setting agendas for adaptation research. Because suitable adaptations differ according to location, sector, and affected parties, and because knowledge about adaptation is widely distributed, adaptation science and technology agendas should be informed by stakeholder interactions. Some guidelines for such interactions are contained in recent NRC reports (NRC, 2008d, 2009b) and a forthcoming report (*ACC: Informing an Effective Response to Climate Change*; NRC, 2010a).

Involving multiple contributors, not just the federal government. Adaptation science and technology advances should be grounded in a partnership among federal, state, and local governments; the private sector and other nongovernmental organizations (NGOs); and the academic research community. This is required because capacities to contribute to topics of interest vary among the partners, research needs to be responsive to the decision-making context of all the decision makers, and science and technology advances should inform voluntary autonomous adaptation actions as well as government program interventions (see below). Mechanisms for effective science and technology partnerships should be developed in a collaborative way.

Co-evolution of science and experience. Because (1) adaptation planning and some adaptation actions are already under way, (2) in many cases adaptation implementation should not wait for years until the knowledge base is strengthened through new research, and (3) current experience is a uniquely valuable source of empirical evidence about what and how adaptation works, it is vitally important to link ongoing science and technology advances with ongoing adaptation experience-building. One important step should be to ensure monitoring and database development to capture and disseminate adaptation experiences (without imposing undue burdens on adaptation decision makers and implementers).

Attention to autonomous adaptation as well as planned adaptation. Finally, it is absolutely essential to avoid an assumption that adaptation only happens because of direct government programs. In many contexts, individual decision makers, from firms to families, are already considering adaptations to stresses associated with climate variability and change. The greater the share of adaptations that can be handled in this way, the more likely adaptation is to be both effective and affordable. A very high priority is to improve our understanding of how to promote, facilitate, and support autonomous adaptation as an alterative or supplement to planned adaptation. One source, for example, suggests that voluntary grassroots action can be encouraged by a combination of significant local control over decisions about priorities, an increased awareness of the risks associated with climate change impacts and the potential benefits of adaptation, and access to a diverse portfolio of technological and institutional alternatives, some of which might not be currently available to some decision makers (Kates and Wilbanks, 2003).

Current Models to Illustrate Options

A number of models for organizing and funding adaptation science and technology are available to illustrate options. The most familiar model in the United States is to mobilize a multi-agency collaboration, depending mainly on individual agency programs but with a provision for multi-agency collaboration. The USGCRP is an example of such an approach, although it has not notably advanced science and technology for adaptation. Australia has responded to strong national concerns about climate change impacts by establishing a National Climate Change Adaptation Research Facility to play a lead role, although it will not be the only source of science and technology advances for adaptation. A solution somewhere between these two approaches might involve assigning lead roles to particular agencies for advancing adaptation science and technology for different sectors, although a coherent national effort would require support for oversight, coordination, and cross-cutting research as well. Finally, an additional example is offered by the United Kingdom Climate Impacts Programme (UKCIP), previously described as a boundary organization, which leads the effort in science and technology development for adaptation in the United Kingdom.

CONCLUSIONS

A lack of serious commitment by the United States to adaptation to climate change has led to an inadequate research effort to provide the science and technology to support appropriate and effective adaptation decisions. Advances in science and technology are needed in the following areas: to support adaptation analysis and assessment, to identify and develop adaptation options, and to strengthen adaptation management and implementation. Many of these advances are needed as quickly as possible to inform such issues as: thresholds or tipping points for climate change impacts that may exceed the limits of adaptation; prospects and approaches for encouraging voluntary relocation from high-vulnerability areas; and relationships between climate change adaptation and issues of resilience and sustainability in a context of multiple threats, stresses, and opportunities. Regarding other high priorities for science and technology advances, see the many challenges listed in this chapter. Adaptation faces challenges not only related to direct changes in climate but also in adapting to the actions people take in responding to climate change (including both GHG emissions-reduction actions and adaptation actions).

> **Conclusion:** In order to strengthen America's choices for adapting to impacts of climate change, science and technology advances are needed in the following areas: adaptation analysis and assessment, adaptation option identification and development, and adaptation management and implementation.

> **Conclusion:** To provide a reliable foundation for adapting to impacts of climate change, in a larger context of sustainability and as a key component of a cross-agency climate change research program (*ACC: Advancing the Science of Climate Change*; NRC, 2010b), the nation needs a significant national strategy and program for climate change adaptation research and development. A shared partnership between the federal government, other levels of government, the private sector and other NGOs, and the academic research community would be the most effective way to achieve this outcome.

Conclusion: Studies of autonomous adaptation as well as planned adaptation are needed, along with monitoring and learning from ongoing experiences with adaptation in practice. A national program could prioritize these needs and also expedite advances in adaptation science and technology that have promise in reducing critical national and regional vulnerabilities to climate change impacts in the coming decades.

Conclusions and Recommendations

Because impacts of climate change are already being observed in the United States and elsewhere in the world, and because impacts will increase in severity even if greenhouse gas (GHG) emissions are reduced substantially in the near term, the United States must improve its ability to adapt to impacts of climate change. Concerns about these impacts are generating increasing interest in adaptation. A wide variety of potential actions that might be taken by individuals, sectors, cities, and states are being discussed—in some cases without sufficient information about the options that are available (GAO, 2009a,b).

Impacts of climate change have the potential to affect all sectors of human and natural systems, depending on the geographic region (Chapter 2), as changes in climate conditions interact with other factors that shape vulnerabilities. The magnitude and rate of future impacts will be shaped significantly by U.S. and global actions to limit emissions (Chapter 3), as well as how the natural Earth system reacts to the resulting emissions trajectory. This means that the magnitude of risks from impacts of climate change involves a great deal of uncertainty. But the certainty of future impacts, and the high likelihood that some of the impacts have a potential to be disruptive to valued human and natural systems, tells us that adaptive responses are unavoidable. The fundamental question is whether we should, as a nation, act proactively to anticipate the impacts of climate change and mobilize to reduce their effects or simply prepare ourselves to react as the impacts arrive.

It is the judgment of this panel that anticipatory adaptation to climate change is a highly desirable risk-management strategy for the United States. Such a strategy offers potentials to reduce costs of current and future climate change impacts by realizing and supporting adaptation capacities across different levels of government, different sectors of the economy, and different populations and environments, and by providing resources, coordination, and assistance in ensuring that a wide range of distributed actions are mutually supportive. Placed in a larger context of sustainable development, climate change adaptation can contribute to a coherent and efficient national response to climate change that encourages linkages and partnerships across boundaries between different types of institutions in our society.

The challenge, however, is considerable. We do not have an institutional infrastructure designed to facilitate an effective approach to adaptation challenges across this country. We do not know enough about adaptation approaches that are available across scales, sectors, and parts of the population. But without a coordinated national approach to adaptation, informed by improved information about our choices, we are unlikely to cope with the impacts of climate change in ways that avoid disruption to society, economy, and ecosystem.

This chapter summarizes the panel's findings and recommendations regarding the need for a national climate change adaptation effort. It emphasizes the term "national" rather than "federal" because adaptation is inherently diverse and disaggregated. Adaptation options to respond to observed and projected climate change impacts (Chapter 2) are immensely diverse; choosing "how" and "when" to adapt from a long list of possible options (Chapter 3) requires careful evaluation of the socioeconomic context, the vulnerability of the sector or region, the resources available, and the scale at which the impact is likely to be felt. There is no one-size-fits-all adaptation option for a particular climate impact across the nation; instead, decision makers within each level of government, within each economic sector, and within civil society need to weigh the many tradeoffs between the available adaptation choices. Most of the decisions about how and when to implement adaptation options will require local input, and in many (if not most) cases, adaptation projects will occur at the local level. The first step in this decision-making process is to better understand the existing vulnerabilities and to consider possible adaptation strategies and options.

Recommendation 1: All decision makers—within national, state, tribal, and local agencies and institutions, in the private sector, and nongovernmental organizations (NGOs)—should identify their vulnerabilities to climate change impacts and the short- and longer-term adaptation options that could increase their resilience to current and projected impacts.

Chapter 4 provides an illustrative approach to such a planning and decision-making process, based on efforts already under way in many cities and states in the United States, and Chapter 5 considers roles and contributions of different members of the American climate change action family. Chapter 6 summarizes how America's climate choices regarding adaptation relate to international contexts, and Chapter 7 summarizes science and technology needs outlined in the preceding chapters.

OVERCOMING ADAPTATION CHALLENGES AND IMPEDIMENTS REQUIRES A COMPREHENSIVE STRATEGY

As indicated above, the panel concludes that realizing America's potential to reduce effects of climate change impacts requires a comprehensive and anticipatory response strategy, bringing together the wide variety of scales, sectors, and concerns that are characteristic of America's approach to solving complex problems. Challenges that call for a comprehensive approach, which might take the form of a national adaptation plan, include the following:

1. *Scales of impacts and resources are often mismatched.*
   Although adaptation has to be implemented at the local and regional scales, some climate change impacts such as sea level rise will exceed the adaptive capacity available at those scales. Many U.S. institutions at virtually every scale lack the mandate, the resources, and/or the professional capacity to select and implement climate change adaptations that will reduce risk sufficiently, even when these adaptation actions are urgently needed. New institutions and bridging organizations will be required to facilitate the communication and integrated planning efforts needed to address complex problems.
2. *Current resource management systems are often based on outdated assumptions.*
   Existing management systems are most often designed around an assumption that the natural environment is essentially stationary—an expectation that future conditions will vary within historic bounds or around a constant average. These assumptions are no longer tenable given the changes already being observed (Milly et al., 2008).
3. *An adaptation option for one sector can put new pressures on another sector.*
   Certain adaptation decisions might adversely impact other sectors, neighboring states, or regions, resulting in a patchwork of actions that may create as many problems as they solve. Because of the projected decrease in snowpack in the western mountains, for example, building reservoirs to increase water storage capacity might help ease the region's water shortages. Yet these actions could also decrease sediment flows to the coast, increasing the problem of coastal erosion and the vulnerability of coastal infrastructure to sea level rise. Conflicting mandates within federal and state agencies managing these sectors make it difficult to align such competing goals to meet the complex interconnected adaptation challenge.
4. *Some adaptation actions are difficult to implement at the state, regional, or local scale due to cost or to the wide range in perception of risk by the public.*

Some proactive adaptation measures, such as moving population and infrastructure away from the coast, are likely to be useful in the longer term as climate change impacts become more visible and damaging. Yet such measures would be extremely difficult to initiate at the local level due to public opposition that hinders proactive actions by politicians and other decision makers and the high initial cost of relocating infrastructure. There is a need to begin planning and investing in studies of such long-term options now in order to ensure that a full array of adaptation options are available when slow-onset impacts manifest in the future.

5. *The nation will not be able to adapt to all the adverse impacts of climate change.* Not all adverse consequences can be avoided through adaptation, although the nation can significantly reduce the extent of damage through proactive actions to avoid, prepare for, and respond to climate change. Establishing adaptation priorities will be required, but such priority decisions will need to be made in the specific decision context. Before priorities can be identified across the nation, consistent methods for conducting vulnerability assessments need to be developed and applied.

Without a well-integrated and coordinated national effort, the United States is currently ill prepared to deal efficiently and effectively with climate challenges. An uncoordinated approach to adaptation in the United States would result in a patchwork of activities that may lead to unintended consequences, conflicting mandates, and potential maladaptations. For this reason, and in keeping with recommendations of the U.S. Conference of Mayors, National Governors Association, and a recent Government Accountability Office report (GAO, 2009b), the panel finds that a national framework is needed to overcome impediments to adaptation, to guide the nation's adaptive response to climate change in a coordinated fashion, and to provide sound advice about how to approach decisions to limit the impacts of climate change.

Again, "national" does not mean "federal government." But, as indicated in Chapter 5, an intermediate multiparty national approach will depend on leadership, a clear strategy, and a centralized coordination mechanism in which the federal government will play a major role, in order to:

- Leverage limited resources;
- Ensure equity in adaptive capacity and investments across needs and geographies;
- Avoid redundant or conflicting projects, mandates, and guidelines;
- Improve understanding of changing conditions;
- Overcome behavior-based limitations to the capacity to adapt; and

- Encourage sharing of information, ideas, and lessons learned.

Although appropriate governance structures and institutions have been identified as a critical component in building adaptive capacity (Adger et al., 2009), U.S. institutions at virtually every scale lack the mandate, the resources, and the professional capacity to select and implement climate change adaptations that will reduce risk sufficiently, even though these adaptation actions are needed (Moser, 2009a). The panel identified a number of approaches to building adaptive capacity in the United States, including encouraging autonomous efforts to adapt led by the private sector, building networks to support adaptation activities within regions and sectors, and establishing a program within the federal government to coordinate and guide adaptation activities across all scales of decision making. In reviewing these options and those taken in support of adaptation efforts in other countries, the panel has concluded that the United States needs to use all of these approaches.

Climate change adaptation represents an entirely new activity for the federal government. The kinds of coordination and support required across the nation and the world will necessitate unprecedented cooperation between agencies and a myriad of interests at the state, local, and international levels. To effectively adapt, the nation will need to facilitate interstate cooperation and coordination across the federal government on adaptation planning, considering such approaches as the following:

1. *Building on and supporting existing efforts and experiences of state and local agencies and partners in the private sector and other NGOs.* The strategy needs to be action- and results-oriented, and should measure progress in terms of improving the nation's adaptive capacity, improving quality of life, and building economic advantages by finding solutions and reducing risks and vulnerabilities to high-priority climate change impacts.

2. *Providing institutional arrangements to link federal incentives (funding, technical assistance, and intergovernmental coordination) with minimum quality standards, and requirements.* Efforts will be needed within regions and within sectors that have historically had limited interaction or actually been in competition with one another. The magnitude and complexity of the adaptation problem requires forging new relationships between the public and private sectors, academia, interest groups, government agencies at all levels, and private citizens. In some cases, it may be most appropriate to develop adaptation plans that are sector-based, such as within the energy industry. In other cases, regional plans or programs may prove more effective. The roles and responsibilities of decision makers at multiple scales will need to be defined and then refined over time.

3. *"Mainstreaming" consideration of climate change adaptation into existing federal programs.* Examples of programs where climate adaptation components, including financial and technical assistance, could be incorporated include the "Farm Bill" (and agricultural policies more generally), the National Flood Insurance Program, agency and program authorization bills, the National Environmental Policy Act, the Transportation Reauthorization Act, and the Endangered Species Act.

4. *Identifying an approach that, in exchange for federal financial and technical support, assists states (and jurisdictions within them, as appropriate) in establishing climate adaptation plans that meet minimum standards for federal approval.* Preparations to limit the impacts of both low-probability, high-impact events and high-probability, low-impact events should be addressed in these plans, as well as proposals to mobilize existing resources, programs, and policies for adaptation and to identify areas where new institutions will be required. The plans should identify resource needs for planning as well as for implementation, and potential existing sources of funding.

5. *Focusing on building climate-resilient systems in all public sectors, including land use planning, energy, water and wastewater systems, transportation systems and infrastructure, stormwater systems, utilities, solid waste management systems, public facilities, coastal hazard planning, public safety services, and health and social services.* Plans should provide a flexible framework for setting priorities and coordinating implementation, including regional partnerships, and should ensure strong public participation and nongovernmental and private-sector stakeholder engagement in planning and implementation (see Chapter 4).

Recommendation 2: The executive branch of the federal government should initiate development of a collaborative national adaptation strategy, which might take the form of a national adaptation plan. The strategy (or plan) should be developed in partnership with congressional leaders, selected high-level representatives of relevant federal agencies, states, tribes, business and environmental organizations, and local governments and community leaders.

Development of a national strategy or national plan should incorporate a "bottom-up" approach that builds on and supports existing efforts and experiences at the state and local levels and efforts of partners in the private sector and other NGOs. In particular, the national adaptation strategy should:

- Establish leadership on climate change adaptation at the highest levels of government;
- Establish a durable vision (including goals, principles, and policy frameworks)

for future public policy decision making with respect to adapting to the impacts of climate change;

- Focus on reducing current and future vulnerabilities to climate change impacts, promoting sustainability, and limiting risks in regions and sectors;
- Aim to develop a coordination mechanism with state and local governments, NGOs, tribes, and the private sector;
- Ensure ongoing climate impact and response assessment activities to provide foci for interactions and information production and sharing as a key part of adaptive risk management and multi-institution coordination;
- Consider minimum standards and guidelines for a wide range of adaptation actions, with the expectation that some states, tribes, and local governments will adopt more stringent standards;
- Focus adaptation efforts on long- and short-term benefits, and capitalize on opportunities to adapt now that may become increasingly difficult in the future;
- Encourage private-sector investments and the development of technologies for adaptation solutions;
- Identify a process to reduce barriers to adaptation that currently exist in legislation, such as incentives for maladaptive behavior and agency mandates that conflict with adaptation goals;
- Address serious needs to improve capacities in major institutions, including staff resources in federal government offices and agencies, to connect adaptation knowledge with society's needs;
- Establish a process to set goals for U.S. policy for climate change adaptation in the international arena; and
- Respond to new science and information on a regular basis and promote an adaptive approach in strategic decisions.

A NATIONAL PROGRAM SHOULD BE DEVELOPED TO IMPLEMENT THE NATIONAL ADAPTATION STRATEGY

Because decision-making entities across all sectors and scales of governance need to develop adaptation plans, the national strategy needs to be tied to effective institutional arrangements for implementation that might include such tools as federal incentives (funding, technical assistance, and intergovernmental consistency), standards, requirements, metrics, and coordination mechanisms to avoid conflicts across agencies or jurisdictional mandates. Effective adaptation will also require a mechanism that facilitates learning from the various adaptation efforts.

To promote consistency across federal, regional, state, and local plans and projects, such institutional arrangements need to include mechanisms to ensure that plans, projects, and grants are effectively coordinated. The federal consistency provision of the federal Coastal Zone Management Act (16 U.S.C. § 1456(c)) could be considered as a model for this critical aspect of intergovernmental coordination, where states are authorized to object to any federal activities that are inconsistent with their federally approved and enforceable coastal policies.

Because public awareness of possible climate change impacts and adaptation strategies is inadequate, well-developed engagement is needed that includes ways to train, leverage, expand, and coordinate existing operational capacity within states, regions, sectors, tribes, the private sector, and NGOs. Public education and extension will be important components of adapting to climate change impacts, because effective adaptation measures will require the participation and support of individual citizens and a variety of sectors and decision makers (*ACC: Informing an Effective Response to Climate Change*; NRC, 2010a).

Especially important is the fact that, because there is a lack of information at local scales about future climate change impacts and great uncertainty about the timing of these impacts, approaches need to be developed that promote flexibility in responding to changing conditions—as opposed to a rigid response intended to be permanent. Adaptive management involves learning from past mistakes; recognizing the complexity and the interrelated nature of sectoral interests such as water, agriculture, and energy; and understanding the relationships between adaptation activities and the need to limit GHG emissions. Over time, there will be a need to adapt to our own adaptations (and maladaptations) as well as to our efforts to limit the magnitude of future climate change.

Recommendation 3: Federal, state, and local governments, together with non-governmental partners, should work together to implement a national climate change adaptation program pursuant to the national climate adaptation strategy. The program should:

- Consider guidelines, minimal standards, and review criteria for adaptation planning and implementation;
- Consider a long-term funding mechanism to support climate change adaptation planning and implementation at all levels that is linked to achieving or exceeding federal standards and guidelines;
- Ensure that a consistent methodology is applied in evaluating plans and setting funding priorities;

- Consider mechanisms to avoid conflicts among federal, state, and local plans through a consultation process;
- Mandate the inclusion of climate change adaptation as a key element in existing federal planning requirements (e.g., Hazard Mitigation Assistance, Federal Highway Administration, etc.) and require federal agencies to build adaptation objectives into their operations, budgets, and planning processes and programs;
- Provide incentives for private-sector participation in solution development;
- Develop long-term strategies now that have a long lead time for implementation and require further evaluation (e.g., strategies to limit development in hazard-prone areas);
- Consider short-term incentives for adaptation options that provide clear benefits over the long term that might not otherwise be initiated due to high initial costs; and
- Educate and engage the public concerning climate change impacts and vulnerabilities through coordinated efforts across agencies, levels of government, and the private sector.

Because of the need to continuously develop new approaches, exchange lessons learned across the nation, evaluate efforts, and train decision makers, a critical component of this national program will be an adaptation support service and network. This support service and network will need to be closely coordinated with the national climate service (*ACC: Informing an Effective Response to Climate Change*; NRC, 2010a), as well as the U.S. Global Change Research Program (*ACC: Advancing the Science of Climate Change*; NRC, 2010b). The program's support service should:

- Build a clearinghouse of adaptation services and best practices built on a series of consistent metrics and deliver information, training, and capacity-building services for climate change adaptation and mitigation that are broadly available to government, NGOs, and private-sector interests and that build upon existing extension programs, adaptation networks, and other current outreach capacity; and
- Provide climate monitoring, mapping, and technical assistance to inform governments at all levels and the private sector on climate impacts and vulnerabilities, as well as to evaluate the effectiveness of adaptation activities and ensure that managers of public lands and resources have adequate support for adaptations to protect ecosystem services and critical habitats.

ADAPTATION SHOULD BE SUPPORTED ACROSS THE NATION BY THE DEVELOPMENT OF NEW ADAPTATION SCIENCE AND TECHNOLOGY

To provide a wider range of choices for the national climate adaptation program and its partners throughout the United States, a new and sustained adaptation research effort will be needed. A lack of serious commitment to adaptation efforts has led to inadequate research support to provide the science and technology needed to support appropriate and effective decisions (NRC, 2009a,b, 2010b), and improving this situation should be a high national priority.

Advances in science and technology are needed to support adaptation analysis and assessment, to identify and develop adaptation options, and to strengthen adaptation management and implementation. Many of these advances are needed very quickly to inform such issues as identifying potential thresholds or tipping points for climate change impacts as they relate to limits of adaptation; prospects and approaches for encouraging voluntary relocation from high-vulnerability areas; and climate change adaptation in a context of sustainability that considers multiple threats, stresses, and opportunities. Adaptation will be required not only to address changes in climate conditions but also society's climate change responses, including emissions-limiting actions, adaptation actions, and potential geoengineering options.

Recommendation 4: As part of an integrated climate change research initiative, the federal government should undertake a significant climate change adaptation research effort designed to provide a reliable foundation for adapting to the impacts of climate change in a larger context of sustainability. This initiative should:

- Be designed as a partnership between the federal government, other levels of government, the private sector and other NGOs, and the academic research community;
- Be developed and implemented in coordination with international partners, state and local governments, NGOs, tribes, and the private sector;
- Consider and be responsive to voluntary, independent adaptation as well as planned adaptation;
- Explicitly include monitoring of ongoing experiences with implementing adaptation to build a clearinghouse for "best practices" that allows sharing of lessons learned; and
- Expedite advances in adaptation science and technology that show promise in reducing vulnerabilities to climate change impacts of particular national and regional concern in the coming decades.

GOVERNMENTS AT ALL LEVELS, THE PRIVATE SECTOR, AND NONGOVERNMENTAL ORGANIZATIONS SHOULD INITIATE ADAPTATION PLANNING AND IMPLEMENTATION

As indicated above, a national adaptation strategy should incorporate knowledge, views, and roles of all aspects of the U.S. economy, society, and environment. The panel chose to focus much of its discussion and analysis on federal, state, and local governments, but it also recommends actions on the part of nongovernmental partners in the national effort.

Recommendation 5: Adaptation planning and implementation at the state and tribal levels should be initiated regardless of whether the federal government provides the necessary leadership. States and tribes will need to take a significant leadership and coordination role, especially in areas where cities and other local interests have not yet established adaptation efforts. State and tribal governments should develop and implement climate change adaptation plans to guide policy and coordinate with federal, regional, local, and private-sector efforts pursuant to the national climate adaptation strategy. These plans should consider:

- A comprehensive assessment, in coordination with other jurisdictions, of climate change impacts, vulnerabilities, and adaptation needs in the context of long-term sustainability objectives;
- A requirement that state and tribal agencies build adaptation objectives into their operations, budgets, planning processes, and programs—including the revisions of environmental review guidelines for state and tribal projects to consider adaptation to climate change vulnerabilities;
- Revisions to state and tribal engineering standards to account for current and anticipated future climate changes;
- Provision of incentives for private-sector and NGO participation in solution development;
- Elimination of public subsidies and incentives for maladaptive activities such as development in high-risk areas;
- Support for the design, implementation, and evaluation of early warning and response systems for climate-sensitive health outcomes; and
- Provisions for adequate support (financial and technical) to protect ecosystem services and critical habitats.

Recommendation 6: Local governments should develop and implement climate change adaptation plans pursuant to the national climate adaptation strategy,

in consultation with the broad range of stakeholders in their communities. These plans should consider:

- Including an assessment of (1) vulnerabilities of all municipal infrastructure to climate change impacts; (2) land use plans, ordinances, and codes to identify opportunities to enhance preparedness for climate change impacts; and (3) resource, staffing, and training needs that would be required to build capacity for adaptation to climate change;
- Building adaptation and mitigation objectives into the operations, budgets, and planning processes and programs of cities and other local governments;
- Including a financial assessment of potential adaptation-related infrastructure needs and operating costs and evaluation of the potential impact of adaptation investments on revenues;
- Designing adaptations to reduce vulnerability to climate change impacts as well as to promote sustainability at a regional level;
- Establishing ongoing monitoring and assessment processes as well as goals and principles for future decision making with respect to adapting to the impacts of climate change; and
- Including a public education and engagement component focusing on local climate change impacts and adaptation issues.

Recommendation 7: The private sector, NGOs, and society at large should assess their own vulnerabilities and risks due to climate change and actively engage and partner with the respective governmental adaptation planning efforts to help build the nation's adaptive capacity.

THE UNITED STATES SHOULD PROMOTE ADAPTATION IN AN INTERNATIONAL CONTEXT

In Chapter 6, the panel considers how U.S. choices on adaptation relate to the international context, including the following perspectives:

- Other than a general recognition of the strategic components of adaptation, the conversation about the U.S. role in international adaptation activities is just beginning. Significant policy questions need to be addressed from the perspective of developing a U.S. adaptation program that recognizes the global context.
- If climate change adaptation objectives are integrated into a range of foreign policy, development assistance, and capacity-building efforts, it is likely that the United States will improve its ability to influence a broader range of

outcomes, including economic and national security considerations. There are multiple ways in which both the opportunities and the risks of climate change are linked across the globe.

- The national security community has identified climate change as a significant factor within the strategic landscape. The potential that climate change will contribute to instability, tension, and conflict as well as increased demand for humanitarian relief has been recognized.

- Current institutions do not provide sufficient support for global adaptation at local scales, where adaptation facilities are needed. They also do not provide sufficiently for exploration of innovative partnerships, techniques, and technologies that could support adaptation action, communication, and trust building between the United States and other countries. New institutions are needed to host international conversations about adaptation, limiting GHG emissions, capacity building, science needs, and geoengineering issues on a peer-to-peer basis.

Recommendation 8: The United States should engage as a major player in adaptation activities at the global scale. The United States should support the establishment of a collaborative, sufficiently funded, international adaptation program that can be sustained over time. The program should:

- Support adaptation projects, capacity building, and sustainable development in countries that have high vulnerability to climate change impacts;
- Include innovative mechanisms for engagement and information exchange and build global adaptation networks; and
- Help coordinate the efforts of public, private, and nongovernmental organizations in international adaptation projects.

Recommendation 9: Adaptation objectives should be incorporated into existing U.S. government programs and policies that have international components, such as (1) agriculture, trade policy, and food security; (2) energy policy; (3) transportation policy; (4) international aid and disaster relief; (5) national security; and (6) intellectual property agreements for technology transfer to other countries.

EARLY OPPORTUNITIES FOR SUCCESS

The decision process about investments in adaptation will evolve and new decision needs will emerge in the future as information about climate change impacts improves and experience reveals the effectiveness of various early adaptation efforts.

This does not mean, however, that no actions should be taken now. In the short term, adaptation might consist of incorporating considerations of climate change impacts into many current policies and resource management practices, a process also referred to as "mainstreaming" adaptation into current policies.

Recommendation 10: Federal, state, and local entities and the private sector should take actions now to address current, known climate change impacts and risks and/or to provide effective risk management at a relatively low cost.

In fact, based on the panel's analysis in previous chapters, a number of adaptation options are available that could be implemented in the short term as risk-management strategies in ways that would not only bring significant near-term benefits but also offer the potential for significant long-term benefits at a relatively modest cost. Examples of actions or mainstreaming adaptation that could be implemented to address major pressing needs within the near-term include the following:

National Priorities

- *Initiate* revisions to the National Flood Insurance Program to require that floodplain maps used for federal flood insurance, state and local regulation, disaster planning, and individual warning take future climate change vulnerabilities into account by reflecting projected changes in sea level rise, storm surge, rainfall-runoff intensity, and flood volumes.
- *Revise* federal, state, and professional engineering standards to reflect current and anticipated future climate changes, and require the use of these standards as a condition for federal investments in infrastructure.
- *Incorporate* adaptation requirements into routine planning, permitting, and investment decisions by existing federal, state, and local authorities.
- *Establish* a database of best practices for adaptation in all sectors.

For Federal Programs

- *Coastal.* Strengthen the ability of the Coastal Zone Management program to address climate impacts by increasing support for the development and implementation of state coastal adaptation plans and strategies.
- *Disaster assistance.* Incorporate climate change adaptation considerations into all federally funded post-disaster redevelopment assistance provided to state and local governments.

- *Environmental impact assessment.* Reexamine and revise guidelines (National Environmental Policy Act and state equivalents) to consider climate change impacts, vulnerabilities, and adaptation options as part of the environmental impact analyses.
- *Foreign assistance.* Incorporate adaptation and sustainability objectives into foreign aid planning and assistance, including the Office of Foreign Disaster Assistance and U.S. Agency for International Development.
- *National security.* Assign responsibility for overseeing the impacts of climate change on national security and for adaptations that increase security.

For Selected Sectors

- *Agriculture.* Review current state and federal regulations and incentives to identify existing requirements and practices that serve as disincentives to adaptation, and identify ways to amend these statutes and policies.
- *Ecosystems.* Implement best management practices (e.g., in fisheries, forests, land use, wetlands) to sustain ecosystem services in a changing climate and to incorporate adaptive management principles in natural resource management plans to reduce ecosystem vulnerabilities.
- *Energy supply and use.* Develop a plan of action with private-sector and state and local partners to enhance the resilience of thermal electric power plants and the U.S. energy grid to climate change impacts and to protect or relocate vulnerable coastal energy infrastructures.
- *Human health and society.* Support the design, implementation, and evaluation of early warning and response systems for climate-sensitive health outcomes, including extreme weather events and infectious disease outbreaks.
- *Transportation.* Revise federal, state, and professional engineering standards to reflect current and anticipated future climate changes and require their use as a condition for federal investments in infrastructure; also, incorporate climate change in the planning process.
- *Urban.* Initiate an integrated assessment of urban infrastructure to determine vulnerabilities to climate change impacts and adaptation needs. One approach that vulnerable communities and states might consider is adopting the International Building Code (International Code Council, 2009).
- *Water.* Provide funding, science, and policy support for the collaborative development of regional water management response strategies to address projected changes in water resources and impacts of extreme events.

In conclusion, although the likely magnitude of climate change impacts is indeed daunting, and the stakes are high, there are a large number of adaptation options that should be initiated now because they are relatively inexpensive, low risk, consistent with sustainability principles, and have multiple co-benefits. The recommendations listed above provide a solid framework within which the nation can initiate a national effort to adapt to the impacts of a changing climate. Along with the near-term activities, it is important to consider adaptation to climate change impacts as a process that will require sustained commitment and a durable yet flexible strategy for several decades to come.

References

ACIA (Arctic Climate Impact Assessment). 2005. *Arctic Climate Impact Assessment.* New York: Cambridge University Press.

Adger, W. N., and K. Vincent. 2005. Uncertainty in adaptive capacity. *Comptes Rendus Geoscience* 337(4):399-410.

Adger, W. N., S. Agrawala, M. M. Q. Mirza, C. Conde, K. O'Brien, J. Pulhin, R. Pulwarty, B. Smit, and K. Takahashi. 2007. Assessment of adaptation practices, options, constraints and capacity. Pp. 717-743 in *Climate Change 2007: Impacts, Adaptation and Vulnerability. Contribution of Working Group II to the Fourth Assessment Report of the Intergovernmental Panel on Climate Change,* M. L. Parry, O. F. Canziani, J. P. Palutikof, P. J. van der Linden, and C.E. Hanson, eds. Cambridge, UK: Cambridge University Press.

Adger, W. N., I. Lorenzoni, and K. L. O'Brien, eds. 2009. *Adapting to Climate Change: Thresholds, Values, Governance.* New York: Cambridge University Press.

AGU (American Geophysical Union). 2006. *Hurricanes and the U.S. Gulf Coast: Science and Sustainable Rebuilding.* Washington, DC: AGU.

AIACC (Assessments of Impacts and Adaptations to Climate Change). 2010. *Assessments of Impacts and Adaptations to Climate Change in Multiple Regions and Sectors 2010.* Online. Available at *http://www.aiaccproject.org/about/about.html.* Accessed February 17, 2010.

Anderies, J. M., M. A. Janssen, and E. Ostrom. 2004. A framework to analyze the robustness of social-ecological systems from an institutional perspective. *Ecology and Society* 9(1):18-34.

Andronova, N. G., and M. E. Schlesinger. 2001. Objective estimation of the probability density function for climate sensitivity. *J. Geophys. Res.,* 106(D19):22605-22611.

Armitage, D., F. Berkes, and N. Doubleday, eds. 2007. *Adaptive Co-Management: Collaboration, Learning, and Multi-level Governance.* Vancouver: University of British Columbia Press.

ASC (American Society of Civil Engineers). 2009. 2009 *Report Card for America's Infrastructure.* Reston, VA: American Society of Civil Engineers. Available at *http://www.infrastructurereportcard.org/report-cards.*

ASFPM (Association of State Floodplain Managers). 2007. *National Flood Programs and Policies in Review 2007.* Madison, WI: ASFPM.

AWWA (American Water Works Association). 2009a. New drinking water infrastructure reports = BIG \$ needs. *AWWA Streamlines* 1(7). March 31, 2009. Available at *http://www.awwa.org/publications/StreamlinesCompleteIssue.cfm?ItemNumber=47296.*

AWWA. 2009b/ *Financing Water Infrastructure: A Water Infrastructure Bank and Other Innovations.* Denver, CO: AWWA. Available at *http:www.awwa.org/fles/GovPublic Affairs/PDF/InfastructureBank.pdf.*

Azar, C. 1999. Weight factors in cost-benefit analysis of climate change. *Environmental and Resource Economics* 13:249-268.

Balbus, J. M., and C. Malina. 2009. Identifying vulnerable subpopulations for climate change health effects in the United States. *Journal of Occupational and Environmental Medicine* 51(1):33-37.

Bales, C.F., and R.D. Duke. 2008. Containing climate change—An opportunity for US leadership. *Foreign Affairs* 87(5):78-89.

Bamberger, R. L., and L. Kumins. 2005. *Oil and Gas: Supply Issues After Katrina and Rita.* Library of Congress, Congressional Research Service.

Bang, G., C. B. Froyn, J. Hovi, and F. C. Menz. 2007. The United States and international climate cooperation: International "pull" versus domestic "push." *Energy Policy* 35(2):1282-1291.

Barnaby, W. 2009. Do nations go to war over water? *Nature* 458(7236):282-283.

Barsugli, J. J., K. Nowak, B. Rajagopalan, J. R. Prairie, and B. Harding. 2009. Comment on "When will Lake Mead go dry?" by T. P. Barnett and D. W. Pierce. *Water Resources Research* 45.

Battisti, D. S., and R. L. Naylor. 2009. Historical warnings of future food insecurity with unprecedented seasonal heat. *Science* 32 (5911):240-244.

Beatley, T. 2009. *Planning for Coastal Resilience: Best Practices for Calamitous Times.* Washington, DC: Island Press.

Beggs, P.J. 2004. Impacts of climate change on aeroallergens: past and future. *Clinical and Experimental Allergy* 34:1507-1513.

Bell, M. L., R. Goldberg, C. Hogrefe, P. L. Kinney, K. Knowlton, B. Lynn, J. Rosenthal, C. Rosenzweig, J. A. Patz. 2007. Climate change, ambient ozone, and health in 50 US cities. *Climatic Change* 82:61-76.

Bierbaum, R., ed. 2007. May. *Coping with Climate Change, Proceedings of a National Summit.* Ann Arbor, MI: School of Natural Resources and Environment, University of Michigan.

Blair, D. C. 2009. *Annual Threat Assessment of the Intelligence Community for the Senate Select Committee on Intelligence.* Online. Available at *http://intelligence.senate.gov/090212/blair.pdf.* Accessed February 12, 2009.

BMBF (Bundesministerium für Bildung und Forschung). 2010. *Klimzug—Climate Change in Regions.* Federal Ministry of Education and Research (BMBF, Bundesministerium für Bildung und Forschung). Online. Available at *http://www.klimzug.de/en/index.php.* Accessed October 15, 2010.

Boardman, A. E., D. H. Greenberg, A. R. Vining, and D. L. Weimer. 2001. *Cost-Benefit Analysis: Concepts and Practice,* 2nd ed. Upper Saddle River, NJ: Prentice Hall.

Boykoff, M. T. 2007. From convergence to contention: United States mass media representations of anthropogenic climate change science. *Transactions of the Institute of British Geographers* 32(4):477-489.

Brauman, K. A., G. C. Daily, T. K. Duarte, and H. A. Mooney. 2007. The nature and value of ecosystem services: An overview highlighting hydrologic services. *Annual Review of Environment and Resources* 32:67-98.

Brekke, L. D., J. E. Kiang, J. R. Olsen, R. S. Pulwarty, D. A. Raff, D. P. Turnipseed, R. S. Webb, and K. D. White. 2009. *Climate Change and Water Resources Management: A Federal Perspective.* U.S. Geological Survey Circular 1331. Reston, VA: U.S. Department of the Interior.

Brewer, T. L. 2008. Climate change technology transfer: A new paradigm and policy agenda. *Climate Policy* 8(5):516-526.

Broder, J. 2009. Climate change seen as a threat to national security. *New York Times,* August 9, 1.

Bronen, R. 2008. Alaskan communities' rights and resilience. *Forced Migration Review* (31):30-32.

Bronen, R. 2009. Forced migration of Alaskan indigenous communities due to climate change: Creating a human rights response. In *Linking Environmental Change, Migration & Social Vulnerability,* edited by A. Oliver-Smith. Bonn: UNU Institute for Environmental and Human Security.

Brooks, M. L., C. M. D'Antonio, D. M. Richardson, J. B. Grace, J. E. Keeley, J. M. DiTomaso, R. J. Hobbs, M. Pellant, and D. Pyke. 2004. Effects of invasive alien plants on fire regimes. *Bioscience* 54(7):677-688.

Burke, M. B., E. Miquel, S. Satyanath, J. A. Dykema, and D. B. Lobell. 2009. Warming increases the risk of civil war in Africa. *Proceedings of the National Academy of Sciences of the United States of America* 106:20670-20674.

Burton, I. 2008. Moving forward on adaptation. Chapter 10 in *From Impacts to Adaptation: Canada in a Changing Climate 2007,* edited by D. S. Lemmen. Ottawa, Ontario: Government of Canada.

Busby, J. W. 2007. *Climate Change and National Security: An Agenda for Action.* New York: Council on Foreign Relations.

California Department of Water Resources. 2008. *Managing an Uncertain Future: Climate Change Adaptation Strategies for California's Water.* Sacramento, CA: California Department of Water Resources.

California Energy Commission. 2009. *California Climate Change Programs.* Online. Available at *http://www.climatechange.ca.gov/policies/ca_activities.html.* Accessed October 26, 2009.

California Urban Water Conservation Council. 2008. *Memorandum of Understanding Regarding Urban Water Conservation in California.* Sacramento, CA: California Urban Water Conservation Council.

Carpenter, S. R., and R. Biggs. 2009. Freshwaters: Managing across scales in space and time. In *Principles of Ecosystem Stewardship: Resilience-Based Natural Resource Management in a Changing World,* edited by F. S. Chapin. New York: Springer.

Carpenter, S. R., H. A. Mooney, J. Agard, D. Capistrano, R. S. DeFries, S. Diaz, T. Dietz, A. K. Duraiappah, A. Oteng-Yeboah, H. M. Pereira, C. Perrings, W. V. Reid, J. Sarukhan, R. J. Scholes, and A. Whyte. 2009. Science for managing ecosystem services: Beyond the Millennium Ecosystem Assessment. *Proceedings of the National Academy of Sciences of the United States of America* 106(5):1305-1312.

Carter, T. R., R. N. Jones, X. Lu, S. Bhadwal, C. Conde, L. O. Mearns, B. C. O'Neill, M. D. A. Rounsevell, and M. B. Zurek. 2007. New assessment methods and the characterization of future conditions. Pp. 133-171 in *Climate Change 2007: Impacts, Adaptation and Vulnerability. Contribution of Working Group II to the Fourth Assessment Report of the Intergovernmental Panel on Climate Change,* edited by M. L. Parry, O. F. Canziani, J. P. Palutikof, P. J. van der Linden, and C. E. Hanson. Cambridge, UK: Cambridge University Press.

CCAP (Center for Clean Air Policy), 2009. *Ask the Climate Question: Adapting to Climate Change Impacts in Urban Regions.* Report by the Center for Clean Air Policy Urban Leaders Adaptation Initiative, edited by A. Lowe, J. Foster, and S. Winkelman. June. Washington, DC: CCAP.

CCSP (Climate Change Science Program). 2007. *Effects of Climate Change on Energy Production and Use in the United States.* Synthesis and Assessment Product 4.5. Report by the U.S. Climate Change Science Program and the subcommittee on Global Change Research, edited by T. J. Wilbanks, V. Bhatt, D. E. Bilello, S. R. Bull, J. Ekmann, W. C. Horak, Y. J. Huang, M. D. Levine, M. J. Sale, D. K. Schmalzer, and M. J. Scott. Washington, DC: Department of Energy, Office of Biological & Environmental Research. 160 pp.

———. 2008a. *Impacts of Climate Change and Variability on Transportation Systems and Infrastructure: Gulf Coast Study, Phase I.* Synthesis and Assessment Product 4.7. Report by the U.S. Climate Change Science Program and the Subcommittee on Global Change Research, edited by M. J. Savonis, V. R. Burkett, and J. R. Potter. Washington, DC: Department of Transportation. 445 pp.

———. 2008b. *Preliminary Review of Adaptation Options for Climate-Sensitive Ecosystems and Resources.* Synthesis and Assessment Product 4.4. Report by the U.S. Climate Change Science Program and the Subcommittee on Global Change Research, J. S. Baron, L. A. Joyce, P. Kareiva, B. D. Keller, M. A. Palmer, C. H. Peterson, and J. M. Scott (authors), edited by S. H. Julius and J. M. West. Washington, DC: U.S. Environmental Protection Agency. 873 pp.

———. 2008c. *The Effects of Climate Change on Agriculture, Land Resources, Water Resources, and Biodiversity in the United States.* Synthesis and Assessment Product 4.3. Report by the U.S. Climate Change Science Program and the Subcommittee on Global Change Research, edited by P. Backlund, A. Janetos, D. Schimel, J. Hatfield, K. Boote, P. Fay, L. Hahn, C. Izaurralde, B. A. Kimball, T. Mader, J. Morgan, D. Ort, W. Polley, A. Thomson, D. Wolfe, M. G. Ryan, S. R. Archer, R. Birdsey, C. Dahm, L. Heath, J. Hicke, D. Hollinger, T. Huxman, G. Okin, R. Oren, J. Randerson, W. Schlesinger, D. Lettenmaier, D. Major, L. Poff, S. Running, L. Hansen, D. Inouye, B. P. Kelly, L. Meyerson, B. Peterson, and R. Shaw. Washington, DC: U.S. Department of Agriculture. 362 pp.

———. 2008d. *Uses and Limitations of Observations, Data, Forecasts, and Other Projections in Decision Support for Selected Sectors and Regions.* Synthesis and Assessment Product 5.1. Report by the U.S. Climate Change Science Program and the Subcommittee on Global Change Research. Washington, DC.

———. 2008e. *Weather and Climate Extremes in a Changing Climate. Regions of Focus: North America, Hawaii, Caribbean, and U.S. Pacific Islands.* Synthesis and Assessment Product 3.3. Report by the U.S. Climate Change Science Program and the Subcommittee on Global Change Research, edited by T. R. Karl, G. A. Meehl, C. D. Miller, S. J. Hassol, A. M. Waple, and W. L. Murray. Washington, DC: Department of Commerce, NOAA's National Climatic Data Center. 164 pp.

———. 2008f. *Analyses of the Effects of Global Change on Human Health and Welfare and Human Systems.* Synthesis and Assessment Product 4.6. Report by the U.S. Climate Change Science Program and the Subcommittee on Global Change Research, K. L. Ebi, F. G. Sussman, and T. J. Wilbanks (authors), edited by J. L. Gamble. Washington, DC: U.S. Environmental Protection Agency.

———. 2009a. *Thresholds of Climate Change in Ecosystems.* Synthesis and Assessment Product 4.2. Report by the U.S. Climate Change Science Program and the Subcommittee on Global Change Research, edited by D. B. Fagre, C. W. Charles, C. D. Allen, C. Birkeland, F. S. Chapin III, P. M. Groffman, G. R. Guntenspergen, A. K. Knapp, A. D. McGuire, P. J. Mulholland, D. P. C. Peters, D. D. Roby, and G. Sugihara. Washington, DC: U.S. Geological Survey, Department of the Interior.

———. 2009b. *Coastal Sensitivity to Sea-Level Rise: A Focus on the Mid-Atlantic Region.* Synthesis and Assessment Product 4.1. Report by the U.S. Climate Change Science Program and the Subcommittee on Global Change Research, J. G. Titus, E. K. Anderson, D. R. Cahoon, S. Gill, R. E. Thieler, and J. S. Williams (authors). Washington, DC: U.S. Environmental Protection Agency.

———. 2009c. *Best Practice Approaches for Characterizing, Communicating, and Incorporating Scientific Uncertainty in Decisionmaking*. Synthesis and Assessment Product 5.2, edited by G. Morgan, H. Dowlatabadi, M. Henrion, D. Keith, R. Lempert, S. McBrid, M. Small, and T. Wilbanks. Washington, DC: National Oceanic and Atmospheric Administration.

CH2M Hill. 2009. *Confronting Climate Change: An Early Analysis of Water and Wastewater Adaptation costs*. Prepared for Association of Metropolitan Water Agencies and National Association of Clean Water Agencies. Available at *http://www.amwa.net/galleries/climate-change/ConfrontingClimateChangeOct09.pdf*. Accessed October 15, 2010

Challenor, P. G., R. K. S. Hankin, and R. Marsh. 2006. Towards the probability of rapid climate change. In *Avoiding Dangerous Climate Change*, edited by H. J. Schellnhuber, W. Cramer, N. Nakicenovic, T. Wigley, and G. Yohe. Cambridge, UK: Cambridge University Press.

Chapin, F. S. III, G. P. Kofinas, C. Folke, and M. C. Chapin. 2009. *Principles of Ecosystem Stewardship Resilience-Based Natural Resource Management in a Changing World*. New York, NY: Springer.

Christensen, J.H., B. Hewitson, A. Busuioc, A. Chen, X. Gao, I. Held, R. Jones, R. K. Koli, W.-T. Kwon, R. Laprise, V. M. Rueda, L.Mearns, C. G.Menéndez, J. Räisänen, A. Rinke, A. Sarr, and P.Whetton. 2007. Regional climate projections. Pp. 847-940 in *Climate Change 2007: The Physical Science Basis. Contribution of Working Group I to the Fourth Assessment Report of the Intergovernmental Panel on Climate Change*, edited by S. Solomon, D. Qin, M. Manning, Z. Chen, M. Marquis, K. B. Averyt, M. Tignor, and H. L. Miller. Cambridge, UK: Cambridge University Press.

CIESIN (Center for International Earth Science Information Network). 2007. *Low Elevation Coastal Zone (LECZ) Urban-Rural Estimates*, Columbia University. Available at *http://sedac.ciesin.columbia.edu/gpw/lecz.jsp*. Accessed October 15, 2010.

Clark, William C., Jill Jaeger, Robert Corell, Roger Kasperson, James J. McCarthy, David Cash, Stewart J. Cohen, Paul Desanker, Nancy M.Dickson, Paul Epstein, David H.Guston, J. Michael Hall, Carlo Jaeger, Anthony Janetos, Neil Leary, Marc A. Levy, Amy Luers, Michael MacCracken, Jerry Melillo, Richard Moss, Joanne M. Nigg, Martin L. Parry, Edward A. Parson, Jesse C. Ribot, Hans Joachim Schellnhuber, Daniel P. Schrag, George A. Seielstad, Eileen Shea, Coleen Vogel, and Thomas J. Wilbanks. 2000. Assessing vulnerability to global environmental risks. Report of the Workshop on Vulnerability to Global Environmental Change: Challenges for Research, Assessment and Decision Making. 22-25 May, Airlie House, Warrenton, Virginia. Research and Assessment Systems for Sustainability Program Discussion Paper 2000-12. Cambridge, MA: Environment and Natural Resources Program, Belfer Center for Science and International Affairs (BCSIA), Kennedy School of Government, Harvard University.

CNA (Center for Naval Analysis). 2007. *National Security and the Threat of Climate Change*. Alexandria, VA: CNA Corporation.

Colten, C. E. 2009. *Perilous Place, Powerful Storms: Hurricane Protection in Coastal Louisiana*. Jackson, MS: University Press of Mississippi.

Commonwealth of Australia. 2009. *Department of Climate Change Corporate Plan 2009-10*. Barton, ACT: Australian Government Department of Climate Change.

Confalonieri, U., B. Menne, R. Akhtar, K.L. Ebi, M. Hauengue, R.S. Kovats, B. Revich and A. Woodward. 2007. Human health. In *Climate Change 2007: Impacts, Adaptation and Vulnerability*, edited by M. L. Parry, O. F. Canziani, J. P. Palutikof, P. J. van der Linden, and C. E. Hanson. Cambridge, UK: Cambridge University Press.

Cooley, H., J. Christian-Smith, P. H. Gleick, L. Allen, and M. Cohen. 2009. *Understanding and Reducing the Risks of Climate Change for Transboundary Waters*. Oakland, CA: Pacific Institute.

Covington, W.W., and M.M.Moore. 1994. Southwestern Ponderosa forest structure—Changes since Euro-American settlement. *Journal of Forestry* 92(1):39-47.

Cruce, T. L. 2009. *Adaptation Planning—What U.S. States and Localities Are Doing*. Prepared for the Pew Center on Global Climate Change, November 2007 (updated August 2009). Available at *http://www.pewclimate.org/publications/workingpaper/adaptation-planning-what-us-states-and-localities-are-doing*. Accessed October 15, 2010 .

CSO (Coastal States Organization). 2007. *The Role of Coastal Zone Management Programs in Adaptation to Climate Change: Final Report of the CSO Climate Change Work Group*. Washington, DC: CSO.

———. 2008. *The Role of Coastal Zone Management Programs in Adaptation to Climate Change: Second Annual Report of the Coastal States Organization's Climate Change Work Group*. Washington, DC: CSO.

Cutter, S. L., L. A. Johnson, C. Finch, and M. Berry. 2007. The U.S. hurricane coasts: Increasingly vulnerable? *Environment* 49(7):8-20.

Daily, G. C., ed. 1997. *Nature's Services: Societal Dependence on Natural Ecosystems.* Washington, DC: Island Press.

de Coninck, H., C. Fischer, R. G. Newell, and T. Ueno. 2008. International technology-oriented agreements to address climate change. *Energy Policy* 36(1):335-356.

de Wit, M., and J. Stankiewicz. 2006. Changes in surface water supply across Africa with predicted climate change. *Science* 311(5769):1917-1921.

Dematte, J. E., K. O'Mara, J. Buescher, C. G. Whitney, S. Forsythe, T. McNamee, R. B. Adiga, and I. M. Ndukwu. 1998. Near-fatal heat stroke during the 1995 heat wave in Chicago. *Annals of Internal Medicine* 129(3):173-181.

Denman, K. L., G. Brasseur, A. Chidthaisong, P. Ciais, P. M. Cox, R. E. Dickinson, D. Hauglustaine, C. Heinze, E. Holland, D. Jacob, U. Lohmann, S Ramachandran, P. L. da Silva Dias, S. C. Wofsy and X. Zhang. 2007. Couplings between changes in the climate system and biogeochemistry. In *Climate Change 2007: The Physical Science Basis. Contribution of Working Group I to the Fourth Assessment Report of the Intergovernmental Panel on Climate Change*, edited by S. Solomon, D. Qin, M. Manning, Z. Chen, M. Marquis, K. B. Averyt, M.Tignor and H. L. Miller. Cambridge, UK: Cambridge University Press.

Department of Communities and Local Government. 2009. *Multi-Criteria Analysis: A Manual.* London: Department of Communities and Local Government.

Diaz, S., J. Fargione, F. S. Chapin, and D. Tilman. 2006. Biodiversity loss threatens human well-being. *PLoS Biology* 4(8):1300-1305.

DOD (Department of Defense). 2006. *Quadrennial Review Report.* Washington, DC: DOD.

DOI (Department of the Interior). 2008a. *Task Force on Climate Change. An Integrated DOI Science Plan for Addressing the Effects of Climate Change on Natural Systems.* Report of the Subcommittee on Science. Washington, DC: DOI.

_____. 2008b. *DOI Climate Change Task Force. An Analysis of Climate Change Impacts and Options Relevant to the Department of the Interior's Managed Lands and Waters.* Report of the Subcommittee on Land and Water Management. Washington, DC: DOI.

_____. 2009. *Order No. 3289: Addressing the Impacts of Climate Change on America's Water, Land, and Other Natural and Cultural Resources.* September 14. Washington, DC: DOI.

DOS (Department of State). 2009. *Quadrennial Diplomacy and Development Review.* Washington, DC: DOS.

Downing, T., and P. Watkiss. 2003. The marginal social costs of carbon in policy making: Applications, uncertainty and a possible risk based approach. In *DEFRA International Seminar on the Social Costs of Carbon.* London.

Easterling, W. E. 1996. Adapting North American agriculture to climate change in review. *Agricultural and Forest Meteorology* 80(1):1-53.

Easterling, W. E., P. K. Aggarwal, P. Batima, K. M. Brander, L. Erda, S. M. Howden, A. Kirilenko, J. Morton, J.-F. Soussana, J. Schmidhuber, and F. N. Tubiello. 2007. Food, fibre and forest products. Pp. 273-313 in *Climate Change 2007: Impacts, Adaptation and Vulnerability. Contribution of Working Group II to the Fourth Assessment Report of the Intergovernmental Panel on Climate Change*, M. L. Parry, O. F. Canziani, J. P. Palutikof, P. J. van der Linden, and C. E. Hanson, eds. Cambridge, UK: Cambridge University Press.

Ebi, K. L., and G. A. Meehl. 2007. The heat is on: Climate change & heatwaves in the Midwest. In *Regional Impacts of Climate Change: Four Case Studies in the United States*, K. L. Ebi, G. A. Meehl, D. Bachelet, R. R. Twilley, and D. Boesch, eds. Pew Center on Global Climate Change, Arlington, VA.

Ebi, K. L., T. J. Teisberg, L. S. Kalkstein, L. Robinson, and R. F. Weiher. 2004. Heat watch/warning systems save lives—Estimated costs and benefits for Philadelphia 1995-98. *Bulletin of the American Meteorological Society* 85(8):1067-1073.

Ebi, K. L., J. Balbus, P. L. Kinney, E. Lipp, D. Mills, M. S. O'Neill and M. Wilson. 2008. Effects of global change and human health. In *Analyses of the Effects of Global Change on Human Health and Welfare and Human Systems.* Washington, DC: U.S. Climate Change Science Program and the Subcommittee on Global Change Research.

EPA (Environmental Protection Agency). 2009. *Climate Ready Estuaries: 2009 Progress Report.* Washington, DC: EPA. Available at *http://www.epa.gov/CRE/downloads/2009-CRE-Progress-Report.pdf.* Accessed October 15, 2010.

Fagre, D. B., C. W. Charles, C. D. Allen, C. Birkeland, F. S. Chapin III, P. M. Groffman, G. R. Guntenspergen, A. K. Knapp, A. D. McGuire, P. J. Mulholland, D. P. C. Peters, D. D. Roby, and G. Sugihara. 2009. *Thresholds of Climate Change in Ecosystems. A Report by the U.S. Climate Change Science Program and the Subcommittee on Global Change Research.* Washington, DC: U.S. Geological Survey, Department of the Interior.

FCOC (Florida Coastal and Ocean Coalition). 2009. *Florida Coastal and Ocean Coalition Policy Report Card.* Tampa, FL: FCOC.

Federal Agencies. 2000. Unified federal policy for a watershed approach to federal land and resource management. *Federal Register* 65(202):62566-62572, October 18.

Feldman, D. L., K. L. Jacobs, G. Garfin, B. Georgakakos, P. Morehouse, R. Restrepo, B. Webb, and D. Yarnal. 2008. Making decision-support information useful, useable, and responsive to decision-maker needs. In *Decision Support Experiments and Evaluations Using Seasonal-to-Interannual Forecasts and Observation Data: A Focus on Water Resources: Synthesis and Assessment Product*, N. K. Ingram and H. M. Beller-Simms, eds. Washington, DC: U.S. Climate Change Science Program.

Feldman, I., and S. Jensen. 2008. Last to the party: The Hill takes up climate change adaptation. *Climate Change, Sustainable Development, and Ecosystems Committee Newsletter* August.

Feldman, I. R., and J. H. Kahan. 2007. Preparing for the day after tomorrow: Frameworks for climate change adaptation. *Sustainable Development Law and Policy* 61(Fall).

Finan, T. J., and D. R. Nelson. 2009. Decentralized planning and climate adaptation: Toward transparent governance. In *Adapting to Climate Change: Thresholds, Values, Governance*, W. N. Adger, I. Lorenzoni, and K. L. O'Brien, eds. Cambridge, UK: Cambridge University Press.

Finucane, M. L., A. Alhakami, P. Slovic, and S. M. Johnson. 2000. The affect heuristic in judgments of risks and benefits. *Journal of Behavioral Decision Making* 13(1):1-17.

Fischhoff, B. 1996. Public values in risk research. *Annals of the American Academy of Political and Social Science* 545:75-84.

Fischlin, A., G. F. Midgley, J. T. Price, R. Leemans, B. Gopal, C. Turley, M. D. A. Rounsevell, O. P. Dube, J. Tarazona, and A. A. Velichko. 2007. Ecosystems, their properties, goods, and services. In *Climate Change 2007: Impacts, Adaptation and Vulnerability. Contribution of Working Group II to the Fourth Assessment Report of the Intergovernmental Panel on Climate Change*, M. L. Parry, O. F. Canziani, J. P. Palutikof, P. J. van der Linden, and C. E. Hanson, eds. Cambridge, UK: Cambridge University Press.

Fisher, B. S., N. Nakicenovic, K. Alfsen, J. Corfee Morlot, F. de la Chesnaye, J.-Ch. Hourcade, K. Jiang, M. Kainuma, E. La Rovere, A. Matysek, A. Rana, K. Riahi, R. Richels, S. Rose, D. van Vuuren, and R. Warren,. 2007. Issues related to mitigation in the long term context, In *Climate Change 2007: Mitigation. Contribution of Working Group III to the Fourth Assessment Report of the Inter-governmental Panel on Climate Change*, B. Metz, O.R. Davidson, P.R. Bosch, R. Dave, L.A. Meyer , eds. Cambridge, UK: Cambridge University Press.

Foley, J. A., R. DeFries, G. P. Asner, C. Barford, G. Bonan, S. R. Carpenter, F. S. Chapin, M. T. Coe, G. C. Daily, H. K. Gibbs, J. H. Helkowski, T. Holloway, E. A. Howard, C. J. Kucharik, C. Monfreda, J. A. Patz, I. C. Prentice, N. Ramankutty, and P. K. Snyder. 2005. Global consequences of land use. *Science* 309(5734):570-574.

Folke, C. 2006. Resilience: The emergence of a perspective for social-ecological systems analyses. *Global Environmental Change* 16(3):253-267.

Frumkin, H., J. Hess, G. Luber, J. Malilay, and M. McGeehin. 2008. Climate change: The public health response. *American Journal of Public Health* 98(3):435-445.

GAO (U.S. Government Accountability Office). 2005a. *21st Century Challenges: Reexamining the Base of the Federal Government.* GAO-05-325SP, Washington, DC: GAO, Feb. 1.

_____. 2005b. *Results-Oriented Government Practices That Can Help Enhance and Sustain Collaboration Among Federal Agencies.* GAO-06-15, Washington, DC: GAO.

_____. 2009a. *Climate Change: Observations on Federal Efforts to Adapt to a Changing Climate.* GAO-09-534T, Washington, DC: GAO, March 25.

_____. 2009b. *Climate Change Adaptation—Strategic Federal Planning Could Help Government Officials Make More Informed Decisions.* GAO-10-113, Report to the Chairman, Select Committee on Energy Independence and Global Warming, House of Representatives.

Garrick, D., M. A. Siebentritt, B. Aylward, C. J. Bauer, and A. Purkey. 2009. Water markets and freshwater ecosystem services: Policy reform and implementation in the Columbia and Murray-Darling Basins. *Ecological Economics* 69(2): 366-379.

Gilbert, N. 2009. Climate refugee fears questioned. *Nature News.* Published online 25 June 2009. Available at *http://www. nature.com/news/2009/090625/full/news.2009.601.html.* Accessed October 15, 2010.

Glantz, M. H. 1988. *Societal Responses to Regional Climatic Change: Forecasting by Analogy.* Boulder, CO: Westview Press.

Glick, P., A. Staudt, and B. Stein. 2009. *A New Era for Conservation: Review of Climate Change Adaptation Literature.* Washington, DC: National Wildlife Federation.

Groves, D. G., and R. J. Lempert. 2007. A new analytic method for finding policy-relevant scenarios. *Global Environmental Change-Human and Policy Dimensions* 17(1):73-85.

Haimes, Y. 1998. *Risk Modeling, Assessment and Management.* New York: John Wiley.

Hannah, L., and L. Hansen. 2005. Designing landscapes and seascapes for change. In *Climate Change and Biodiversity,* T. E. Lovejoy and L. Hannah, eds. New Haven, CT: Yale University Press.

Hansen, L. J., J. L. Biringer, and J. R. Hoffman, eds. 2003. *Buying Time: A User's Manual for Building Resistance and Resilience to Climate Change in Natural Systems.* Washington, DC: World Wildlife Fund Climate Change Program.

Hawking, F., and R. Sutton. 2009. The potential to narrow uncertainty in regional climate predictions. *Bulletin of the American Meteorological Society* 90:1095-1107.

Heinz Center. 2000. *The Hidden Costs of Coastal Hazards.* Washington, DC: Island Press.

Heller, N. E., and E. S. Zavaleta. 2009. Biodiversity management in the face of climate change: A review of 22 years of recommendations. *Biological Conservation* 142:14-32.

Hinzman, L. D., N. D. Bettez, W. R. Bolton, F. S. Chapin, M. B. Dyurgerov, C. L. Fastie, B. Griffith, R. D. Hollister, A. Hope, H. P. Huntington, A. M. Jensen, G. J. Jia, T. Jorgenson, D. L. Kane, D. R. Klein, G. Kofinas, A. H. Lynch, A. H. Lloyd, A. D. McGuire, F. E. Nelson, W. C. Oechel, T. E. Osterkamp, C. H. Racine, V. E. Romanovsky, R. S. Stone, D. A. Stow, M. Sturm, C. E. Tweedie, G. L. Vourlitis, M. D. Walker, D. A. Walker, P. J. Webber, J. M. Welker, K. Winker, and K. Yoshikawa. 2005. Evidence and implications of recent climate change in northern Alaska and other arctic regions. *Climatic Change* 72(3):251-298.

Hobbs, R. J., D. N. Cole, E. S. Yung, G. H. Zavaleta, F. S. Aplet, P. B. Chapin III, D. J. Landres, N. L. Parsons, P. S. Stephenson, and D. M. White. 2009. Guiding concepts for parks and wilderness areas stewardship in an era of global environmental change. *Frontiers in Ecology and the Environment.* E-View pre-print, Dec 2, 2009. Available at *http://www.treesearch. fs.fed.us/pubs/34151.* Accessed October 15, 2010.

Holling, C. S. 1978. *Adaptive Environmental Assessment and Management.* Laxenburg, Austria: International Institute for Applied Systems Analysis.

Howden, M. 2009. Australia's approach to adaptation. Presentation to the America's Climate Choices Panel on Adapting to the Impacts of Climate Change, May 4. Commonwealth Scientific and Industrial Research Organisation.

Howden, S. M., J.-F. Soussana, F. N. Tubiello, N. Chhetri, M. Dunlop, and H. Meinke. 2007. Adapting agriculture to climate change. *Proceedings of the National Academy of Sciences of the United States of America* 104(50): 19691-19696.

Huber, O., R. Wider, and O. W. Huber. 1997. Active information search and complete information presentation in naturalistic risky decision tasks. *Acta Psychologica* 95(1):15-29.

Huntington, H. P. 2000. Native observations capture impacts of sea ice changes. *Witness the Arctic* 8(1):1-2.

Hurteau, M. D., G. W. Koch, and B. A. Hungate. 2008. Carbon protection and fire risk reduction: Toward a full accounting of forest carbon offsets. *Frontiers in Ecology and the Environment* 6(9):493-498.

IAW (Immediate Action Workgroup). 2009. Recommendations to the Governor's Sub-Cabinet on Climate Change. Juneau.

ICLEI (Local Governments for Sustainability). 2007. *Keene New Hampshire Climate Change Adaptation Plan: Summary Report.* Oakland, CA: ICLEI.

International Code Council. 2009. *International Building Code.* International Code Council, Inc. Cengage Learning, 676 pp. Available at *http://publicecodes.citation.com/icod/ibc/2009/index.htm?bu=IC-P-2009-000001&bu2=IC-P-2009-000019.* Accessed October 15, 2010.

IOM (Institutional Organization for Migration). 2008. *Climate Change and Migration: Improving Methodologies to Estimate Flows.* IOM Migration Research Series, 33. IOM: Geneva.

IPCC (Intergovernmental Panel on Climate Change). 2001a. *Climate Change 2001: Impacts, Adaptation & Vulnerability. Contribution of Working Group II to the Third Assessment Report of the Intergovernmental Panel on Climate Change*. New York: Cambridge University Press.

———. 2001b. *Climate Change 2001: Synthesis Report. A Contribution of Working Groups I, II and III to the Third Assessment Report of the Intergovernmental Panel on Climate Change*. New York: Cambridge University Press.

———. 2007a. *Climate Change 2007: Impacts, Adaptation and Vulnerability. Contribution of Working Group II to the Fourth Assessment Report of the Intergovernmental Panel on Climate Change*, M. L. Parry, O. F. Canziani, J. P. Palutikof, P. J. van der Linden, and C. E. Hanson, eds. Cambridge, UK: Cambridge University Press.

———. 2007b. *Climate Change 2007: The Physical Science Basis. Contribution of Working Group I to the Fourth Assessment Report of the Intergovernmental Panel on Climate Change*, S. Solomon, D. Qin, M. Manning, Z. Chen, M. Marquis, K. B. Averyt, M. Tignor, and H. L. Miller, eds. Cambridge, UK: Cambridge University Press.

———. 2007c. *Climate Change 2007: Synthesis Report. Contribution of Working Groups I, II and III to the Fourth Assessment Report of the Intergovernmental Panel on Climate Change*. Core Writing Team, R. K. Pachauri, and A. Reisinger, eds. Geneva, Switzerland: IPCC, 104 pp.

Jackson, R., and K. N. Shields. 2008. Preparing the U.S. health community for climate change. *Annual Review of Public Health* 29:57-73.

Jacobs, K., G. Garfin, and M. Lenart. 2005. More than just talk: Connecting science and decisionmaking. *Environment* 47(9):6-21.

Jones, R., and L. Mearns. 2005. Assessing future climate risks. In *Adaptation and Policy Framework for Climate Change: Developing Strategies, Policies and Measures*, B. Lim and E. Spanger-Siegfried, eds. Cambridge: Cambridge University Press.

Kalkstein, L. S., P. F. Jamason, J. S. Greene, J. Libby, and L. Robinson. 1996. The Philadelphia hot weather-health watch warning system: Development and application, summer 1995. *Bulletin of the American Meteorological Society* 77(7):1519-1528.

Kalkstein, L. S., J. S. Greene, D. M. Mills, A. D. Perrin, J. P. Samenow, and J. C. Cohen. 2008. Analog European heat waves for US cities to analyze impacts on heat-related mortality. *Bulletin of the American Meteorological Society* 89(1):75-85.

Kates, R. W. 2007. The retreat from the coast. *Environment* 49(7): C2-C4.

Kates, R. W., and T. J. Wilbanks. 2003. Making the global local: Responding to climate change concerns from the ground up. *Environment* 45(3):12-23.

Kates, R. W., C. E. Colten, S. Laska, and S. P. Leatherman. 2006. Reconstruction of New Orleans after Hurricane Katrina: A research perspective. *Proceedings of the National Academy of Sciences of the United States of America* 103(40):14653-14660.

Keeney, R. L., and H. Raiffa. 1976. *Decisions with Multiple Objectives: Preferences and Value Tradeoffs*. New York: John Wiley.

Keiter, R. B., and M. S. Boyce. 1991. *The Greater Yellowstone Ecosystem: Redefining America's Wilderness Heritage*. New Haven, CT: Yale University Press.

Khagram, S., and S. Ali. 2006. Environment and security. *Annual Review of Environment and Resources* 31:395-411.

Kilbourne, E. M. 1997. Heat waves and hot environments. In *The Public Health Consequences of Disasters*. New York: Oxford University Press.

King County. 2007. *2007 King County Climate Plan*. February. Available at *http://your.kingcounty.gov/exec/news/2007/pdf/climateplan.pdf*. Accessed October 15, 2010.

Kinney, P.L. 2008. Climate Change, Air Quality, and Human Health. *American Journal of Preventive Medicine* 35(5):459-467.

Klein, R. J. T., and A. Möhner. 2009. Governance limits to effective global financial support for adaptation. In *Adapting to Climate Change: Thresholds, Values, Governance*, W. N. Adger, I. Lorenzoni, and K. L. O'Brien, eds. New York: Cambridge University Press.

Kovats, R. S., and S. Hajat. 2008. Heat stress and public health: A critical review. *Annual Review of Public Health*. 29: 41-55.

Krantz, D. H., and H. C. Kunreuther. 2007. Goals and plans in decision making. *Judgment and Decision Making Journal* 2(3):137-168.

Kundzewicz, Z. W., L. J. Mata, N. W. Arnell, P. Döll, P. Kabat, B. Jiménez, K. A. Miller, T. Oki, Z. Sen, and I. A. Shiklomanov. 2007. Freshwater resources and their management. Pp. 173-210 in *Climate Change 2007: Impacts, Adaptation and Vulnerability. Contribution of Working Group II to the Fourth Assessment Report of the Intergovernmental Panel on Climate Change*, M. L. Parry, O. F. Canziani, J. P. Palutikof, P. J. van der Linden, and C. E. Hanson, eds. Cambridge, UK: Cambridge University Press.

Kunreuther, H., and E. Michel-Kerjan. 2009. From market to government failure in insuring U.S. natural catastrophes: How long term contracts help. In *Private Markets and Public Insurance Programs*, J. Brown, ed. Washington DC: American Enterprise Institute.

Lawler, J. J., T. H. Tear, C. Pyke, R. Shaw, P. Gonzalez, P. Kareiva, L. Hansen, L. Hannah, K. Klausmeyer, A. Aldous, C. Bienz, and S. Pearsall. 2009. Resource management in a changing and uncertain climate. *Frontiers in Ecology and the Environment* 7 (online e-view) Available at *http://depts.washington.edu/landecol/PDFS/gcc-managers-review.pdf*. Accessed on October 15, 2010.

Leiserowitz, A. A. 2005. American risk perceptions: Is climate change dangerous? *Risk Analysis* 25(6):1433-1442.

Lempert, R., N. Nakicenovic, D. Sarewitz, and M. Schlesinger. 2004. Characterizing climate-change uncertainties for decision-makers. An editorial essay. *Climatic Change* 65(1-2):1-9.

Lenart, M. 2006. Collaborative stewardship to prevent wildfires. *Environment* 48(7):8-21.

Lenton, T. M., H. Held, E. Kriegler, J. M. Hall, W. Lucht, S. Rahmstorf, and H. J. Schellnhuber. 2008. Tipping points in the Earth's climate system. *Proceedings of the National Academy of Sciences of the United States of America* 105(6):1786-1793.

Levy, D. L., and S. Rothenberg. 2002. Heterogeneity and change in environmental strategy: Technological and political responses to climate change in the global automobile industry. Pp. 173-193 in *Organizations, Policy, and the Natural Environment: Institutional and Strategic Perspectives*, A. J. Hoffman and M. J. Ventresca, eds. Stanford: Stanford University Press.

Lloyd's of London. 2008. *Coastal Communities and Climate Change*. London: Llloyd's of London.

Loewenstein, G., and D. Prelec. 1992. Anomalies in intertemporal choice—Evidence and an interpretation. *Quarterly Journal of Economics* 107(2):573-597.

Loewenstein, G. F., E. U. Weber, C. K. Hsee, and N. Welch. 2001. Risk as feelings. *Psychological Bulletin* 127(2):267-286.

Loomis, J. B. 1993. *Integrated Public Lands Management: Principles and Applications to National Forests, Parks, Wildlife Refuges, and BLM Lands*. New York: Columbia University Press.

Ludwig, F., P. Kabat, H. van Schaik, and M. van der Valk, eds. 2009. *Climate Change Adaptation in the Water Sector*. London: Earthscan.

MAB (Military Advisory Board). 2007. *National Security and the Threat of Climate Change*. The CNA Corporation, *http://SecurityAndClimate.cna.org*, Accessed October 15, 2010. 63 pp.

Marino, E. 2009. Immanent threats, impossible moves, and unlikely prestige: Understanding the struggle for local control as a means toward sustainability. In *Linking Environmental Change, Migration and Social Vulnerability*. Bonn, Germany: United Nations University.

McKenzie, D., Z. Gedalof, D. L. Peterson, and P. Mote. 2004. Climatic change, wildfire, and conservation. *Conservation Biology* 18(4):890-902.

MEA (Millennium Ecosystem Assessment). 2005. *Ecosystems and Human Well-Being: Synthesis*. Washington, DC: Island Press.

Mearns, L. O., and M. Hulme. 2001. Climate scenario development. Pp. 739-768 in Climate Change 2001: The Scientific Basis, *Contribution of Working Group II to the Third Assessment Report of the Intergovernmental Panel on Climate Change*. Cambridge, UK: Cambridge University Press.

Meehl, G. A., T. F. Stocker, W. D. Collins, A. T. Friedlingstein, A. T. Gaye, J. M. Gregory, A. Kitoh, R. Knutti, J. M. Murphy, A. Noda, S. Raper, I. G. Watterson, A. J. Weaver, and Z. Zhao. 2007. Global climate projections. Pp 747-845 in *Climate Change 2007: The Physical Science Basis. Contribution of Working Group I to the Fourth Assessment Report of the Intergovernmental Panel on Climate Change*. Cambridge: Cambridge University Press.

Mileti, D., and J. Gailus. 2005. Sustainable development and hazards mitigation in the United States: Disasters by design revisited. *Mitigation and Adaptation Strategies for Global Change* 10(3):491-504.

Millar, C. I., N. L. Stephenson, and S. L. Stephens. 2007. Climate change and forests of the future: Managing in the face of uncertainty. *Ecological Applications* 17(8):2145-2151.

Milly, P. C. D., J. Betancourt, M. Falkenmark, R. M. Hirsch, Z. W. Kundzewicz, D. P. Lettenmaier, and R. J. Stouffer. 2008. Climate change—Stationarity is dead: Whither water management? *Science* 319(5863):573-574.

Mirchandani, H. G., G. McDonald, I. C. Hood, and C. Fonseca. 1996. Heat-related deaths in Philadelphia—1993. *American Journal of Forensic Medicine and Pathology* 17(2):106-108.

Moser, S. 2009a. Whether our levers are long enough and the fulcrum strong? Exploring the soft underbelly of adaptation decisions and actions. Pp. 313-334 in *Adapting to Climate Change: Thresholds, Values, Governance*, W. N. Adger, I. Lorenzoni, and K. O'Brien, eds. Cambridge, UK: Cambridge University Press.

_____. 2009b. *Good Morning America! The Explosive US Awakening to the Need for Adaptation*. Sacramento, CA: California Energy Commission; Charleston, SC: NOAA-Coastal Services Center.

Moser, S., G. Franco, S. Pittiglio, W. Chou, and D. Cayan. 2009. *The Future is Now: An Update on Climate Change Science Impacts and Response Options for California*. Sacramento: California Energy Commission.

Moss, R., and S. H. Schneider. 2000. Uncertainties in the IPCC TAR: Recommendations to lead authors for more consistent assessment and reporting. Pp. 33-51 in *Guidance Papers on the Cross Cutting Issues of the Third Assessment Report of the IPCC*. R. Pachauri, T. Taniguchi, and K. Tanaka, eds. Technical Report. Geneva: World Meteorological Organization.

Moss, R. H., A. L. Brenkert, and E. L. Malone. 2001. *Vulnerability to Climate Change: A Quantitative Approach*. Richland, WA: Pacific Northwest National Laboratory.

Murphy, J. M., D. M. H. Sexton, D. N. Barnett, G. S. Jones, M. J. Webb, M. Collins, and D. A. Stainforth. 2004. Quantification of modeling uncertainties in a large ensemble of climate change simulations. *Nature* 430:768-772.

Murphy, J. M., D. Sexton, G. Jenkins, P. Boorman, B. Booth, K. Brown, R. Clark, M. Collins, G. Harris, and E. Kendon. 2009. Online Climate Change Projections Report. UKCP09 Climate Change Projections. Available at *http://ukclimateprojections. defra.gov.uk/content/view/824/517*. Accessed October 15, 2010.

Naylor, R. L. 2009. Managing food production systems for resilience. In *Principles of Ecosystems Stewardship: Resilience-Based Natural Resource Management in a Changing World*, F. S Chaplin, ed. New York: Springer.

Neumann, J. E., and J. C. Price. 2009. *Adapting to Climate Change, The Public Policy Response: Public Infrastructure*. Washington, DC: Resources for the Future.

Neumann, J. E., G. Yohe, R. Nicholls, and M. Manion. 2000. *Sea-Level Rise and Global Climate Change: A Review of Impacts to U.S. Coasts*. Pew Center on Global Climate Change. Available at *http://www.pewclimate.org/docUploads/env_sealevel. pdf*. Accessed October 15, 2010.

NGA (National Governors Association). 2009. *Policy Position: NR-10. Ocean and Coastal Zone Management (02/25/2009), 10.5 Coastal Adaptation to Climate Change*. Washington, DC: National Governors Association.

Niasse, M. 2005. Climate-induced water conflict risks in West Africa: Recognizing and coping with increasing climate impacts on shared watercourses. In *Human Security and Climate Change: An International Workshop*. Oslo: Global Environmental Change and Human Security Program (GECHS).

NOAA (National Oceanic and Atmospheric Administration). 2006. *Coastal Zone Management Program—Enhancement Grant Assessments and Strategies: Coastal Hazards* Available at *http://coastalmanagement.noaa.gov/issues/docs/ hazards_summary.pdf*. Accessed October 15, 2010.

_____. 2009. *Regional Integrated Sciences and Assessments (RISA) Programs—FY 2009 Information Sheet*. NOAA Climate Program Office. Available at *http://www.cpo.noaa.gov/opportunities/2009/pdf/risa_info.pdf*. Accessed October 15, 2010.

Nordhaus, W. D., and J. Boyer. 2000. *Warming the World: Economic Models of Global Warming*. Cambridge, MA: MIT Press.

NPCC (New York City Panel on Climate Change). 2009. *Climate Risk Information: New York City Panel on Climate Change*. New York: NPCC.

_____. 2010. *Climate Change Adaptation in New York City: Building a Risk Management Response*, C. Rosenzweig and W. Solecki, eds. Prepared for use by the New York City Climate Change Adaptation Task Force. *Annals of the New York Academy of Sciences* 1196: 1-354.

NRC (National Research Council). 1987. *Responding to Changes in Sea Level: Engineering Implications*. Washington, DC: National Academy Press.

_____.1992. *Policy Implications of Greenhouse Warming: Mitigation, Adaptation, and the Science Base*. Washington, DC: National Academy Press.

———. 1996a. *A New Era for Irrigation*. Washington, DC: National Academy Press.

———. 1996b. *Understanding Risk: Informing Decisions in a Democratic Society*. Washington, DC: National Academy Press..

———. 1999. *Human Dimensions of Global Environmental Change: Research Pathways for the Next Decade*. Washington, DC: National Academy Press.

———. 2001a. *Grand Challenges in Environmental Sciences*. Washington, DC: National Academy Press.

———. 2001b. *Improving the Effectiveness of U.S. Climate Modeling*. Washington, DC: National Academy Press.

———. 2002. *Abrupt Climate Change: Inevitable Surprises*. Washington, DC: National Academy Press..

_____. 2004. *Adaptive Management for Water Resources Project Planning*. Washington, DC: The National Academies Press.

———. 2005a. *Decision Making for the Environment: Social and Behavioral Science Research Priorities*. Washington, DC: The National Academies Press.

———. 2005b. *Thinking Strategically: The Appropriate Use of Metrics for the Climate Change Science Program*. Washington, DC: The National Academies Press.

———. 2006. *Drawing Louisiana's New Map: Addressing Land Loss in Coastal Louisiana*. Washington, DC: The National Academies Press.

———. 2007a. *Analysis of Global Change Assessments: Lessons Learned*. Washington, DC: The National Academies Press.

———. 2007b. *Colorado River Basin Water Management: Evaluating and Adjusting to Hydroclimatic Variability*. Committee on the Scientific Bases of Colorado River Basin Water Management, ed. Washington, DC: The National Academies Press.

———. 2007c. *Mitigating Shore Erosion Along Sheltered Coasts*. Washington, DC: The National Academies Press.

———. 2007d. *Evaluating Progress of the U.S. Climate Change Science Program: Methods and Preliminary Results*. Washington, DC: The National Academies Press.

———. 2008a. *Ecological Impacts of Climate Change*. Washington, DC: The National Academies Press.

———. 2008b. *Increasing Capacity for Stewardship of Oceans and Coasts: A Priority for the 21st Century*. Washington, DC: The National Academies Press.

———. 2008c. *Potential Impacts of Climate Change on U.S. Transportation: Special Report 290*. Washington, DC: The National Academies Press.

———. 2008d. *Public Participation in Environmental Assessment and Decision Making*. Washington, DC: The National Academies Press.

———. 2009a. *Informing Decisions in a Changing Climate*. Washington, DC: The National Academies Press.

———. 2009b. *Mapping the Zone: Improving Flood Map Accuracy*. Washington, DC. The National Academies Press.

———. 2009c. *Restructuring Federal Climate Research to Meet the Challenges of Climate Change*. Washington, DC: The National Academies Press.

_____. 2010a. *ACC: Informing an Effective Response to Climate Change*. Washington, DC: The National Academies Press.

_____. 2010b. *ACC: Advancing the Science of Climate Change*. Washington, DC: The National Academies Press.

_____. 2010c. *ACC: Limiting the Magnitude of Future Climate Change*. Washington, DC: The National Academies Press.

_____. 2010d. *Ocean Acidification*. Washington, DC: The National Academies Press.

NROC (Northeast Regional Ocean Council). 2009. *Coastal Hazards Resilience Committee 2009-2010 Workplan*. Available at *http://collaborate.csc.noaa.gov/nroc/Shared%20Documents/NROC%20Committee%20-%20Hazards%20Resilience/HazardsResilience%20WorkPlan%20(v9-21-09).pdf*. Accessed October 15, 2010.

NSTC (National Science and Technology Council). 2008. *Scientific Assessment of the Effects of Global Change on the United States*. Washington, DC: NSTC.

O'Brien, K., S. Eriksen, L. Sygna, and L. O. Naess. 2006. Questioning complacency: Climate change impacts, vulnerability, and adaptation in Norway. *Ambio* 35(2):50-56.

Oladosu, G., and A. Rose. 2007. Income distribution impacts of climate change mitigation policy in the Susquehanna River Basin economy. *Energy Economics* 29(3):520-544.

O'Neill, B. 2005. U.S. Socio-Economic Futures. Manuscript. Laxenburg, Austria: International Institute for Applied Systems Analysis.

Palecki, M. A., S. A. Changnon, and K. E. Kunkel. 2001. The nature and impacts of the July 1999 heat wave in the midwestern United States: Learning from the lessons of 1995. *Bulletin of the American Meteorological Society* 82(7):1353-1367.

Parmesan, C., and H. Galbraith. 2004. *Observed Impacts of Global Climate Change in the U.S.* Washington, DC: Pew Center on Global Climate Change.

Parry, M., N. Arnell, P. Berry, D. Dodman, S. Fankhauser, C. Hope, S. Kovats, R. Nicholls, D. Satterthwaite, R. Tiffin, T. Wheeler. 2009. *Assessing the Costs of Adaptation to Climate Change: A Review of the UNFCCC and Other Recent Estimates.* London: International Institute for Environment and Development and Grantham Institute for Climate Change.

Parson, E., V. Burkett, K. Fisher-Vanden, D. Keith, L. Mearns, H. Pitcher, C. Rosenzweig, and M. Webster. 2007. *Global Change Scenarios: Their Development and Use, 2007. Sub-Report 2.1B of Synthesis and Assessment Product 2.1 by the U.S. Climate Change Science Program and the Subcommittee on Global Change Research.* Washington, DC: Department of Energy, Office of Biological and Environmental Research, p. 106.

Paton, D. 2008. Risk perception and volcanic hazard mitigation: Individual and social perspectives. *Journal of Volcanology and Geothermal Research* 172:179-188.

Patt, A., and S. Dessai. 2005. Communicating uncertainty: Lessons learned and suggestions for climate change assessment. *Geoscience* 337(4):425-441.

Patz, J. A., P. R. Epstein, T. A. Burke, and J. M. Balbus. 1996. Global climate change and emerging infectious diseases. *Journal of the American Medical Association* 275:217-223.

Peters, G. 2009. *Seeds of Terror.* New York: St. Martin's Press.

Pew Center on Global Climate Change. 2009. *Adaptation Plans,* Available at *http://www.pewclimate.org/what_s_being_done/in_the_states/adaptation_map.cfm.* Accessed September 2009.

Porter, D. R. 1997. *Managing Growth in America's Communities.* Washington, DC: Island Press.

Post, E., M. C. Forchhammer, M. S. Bret-Harte, T. V. Callaghan, T. R. Christensen, B. Elberling, A. D. Fox, O. Gilg, D. S. Hik, T. T. Hã,ye, R. A. Ims, E. Jeppesen, D. R. Klein, J. Madsen, A. D. McGuire, S. Rysgaard, D. E. Schindler, I. Stirling, M. P. Tamstorf, N. J. C. Tyler, R. Van Der Wal, J. Welker, P. A. Wookey, N. M. Schmidt, and P. Aastrup. 2009. Ecological dynamics across the arctic associated with recent climate change. *Science* 325(5946):1355-1358.

Presidential Directive. 2009. White House Memorandum on Arctic Region Policy, National Security Presidential Directive and Homeland Security Presidential Directive, NSPD 66/HSPD 25.

Radeloff, V. C., R. B. Hammer, S. I. Stewart, J. S. Fried, S. S. Holcomb, and J. F. McKeefry. 2005. The wildland-urban interface in the United States. *Ecological Applications* 15(3):799-805.

Reeder, T., J. Wicks, L. Lovell, and O. Tarrant. 2009. Protecting London from tidal flooding: Limits to engineering adaptation. In *Adapting to Climate Change: Thresholds, Values, Governance,* W. N. Adger, I. Lorenzoni, and K. L. O'Brien, eds. Cambridge, U.K.: Cambridge University Press.

Reilly, J., F. Tubiello, B. McCarl, and J. Melillo. 2001. Climate change and agriculture in the United States. In *Climate Change Impacts on the United States: The Potential Consequences of Climate Variability and Change, Report for the US Global Change Research Program.* Cambridge: Cambridge University Press.

Richels, R. G., and G. J. Blanford. 2008. The value of technological advances in decarbonizing the U.S. economy. *Energy Economics* 30(6):2930-2946.

Robine, J. M., S. L. K. Cheung, S. Le Roy, H. Van Oyen, C. Griffiths, J. P. Michel, and F. R. Herrmann. 2008. Death toll exceeded 70,000 in Europe during the summer of 2003. *Comptes Rendus Biologies* 331 (2):171-175.

Rockstrom, J., W. Steffen, K. Noone, A. Persson, F. S. Chapin, E. F. Lambin, T. M. Lenton, M. Scheffer, C. Folke, H. J. Schellnhuber, B. Nykvist, C. A. de Wit, T. Hughes, S. van der Leeuw, H. Rodhe, S. Sorlin, P. K. Snyder, R. Costanza, U. Svedin, M. Falkenmark, L. Karlberg, R. W. Corell, V. J. Fabry, J. Hansen, B. Walker, D. Liverman, K. Richardson, P. Crutzen, and J. A. Foley. 2009. A safe operating space for humanity. *Nature* 461(7263):472-475.

Rosenzweig, C., D. Karoly, M. Vicarelli, P. Neofotis, Q. G. Wu, G. Casassa, A. Menzel, T. L. Root, N. Estrella, B. Seguin, P. Tryjanowski, C. Z. Liu, S. Rawlins, and A. Imeson. 2008. Attributing physical and biological impacts to anthropogenic climate change. *Nature* 453(7193):353-320.

Salsich, P. W., and T. J. Tryniecki. 1998. *Land Use Regulation: A Legal Analysis & Practical Application of Land Use Law*. Chicago: Real Property, Probate and Trust Law, American Bar Association.

Schipper, L., and M. Pelling. 2006. Disaster risk, climate change and international development: Scopes for, and challenges to, integration. *Disasters* 30(1):19-38.

Schmidhuber, J., and F. N. Tubiello. 2007. Global food security under climate change. *Proceedings of the National Academy of Sciences of the United States of America* 104(50):19703-19708.

Schneider, S. H., S. Semenov, A. Patwardhan, I. Burton, C. H. D. Magadza, M. Oppenheimer, A. B. Pittock, A. Rahman, J. B. Smith, A. Suarez, and F. Yamin. 2007. Assessing key vulnerabilities and the risk from climate change. Pp. 779-810 in *Climate Change 2007: Impacts, Adaptation and Vulnerability. Contribution of Working Group II to the Fourth Assessment Report of the Intergovernmental Panel on Climate Change*, M. L. Parry, O. F. Canziani, J. P. Palutikof, P. J. van der Linden, and C. E. Hanson, eds. Cambridge, UK: Cambridge University Press.

Schoennagel, T., T. T. Veblen, and W. H. Romme. 2004. The interaction of fire, fuels, and climate across rocky mountain forests. *Bioscience* 54(7):661-676.

Schoennagel, T., C. R. Nelson, D. M. Theobald, G. C. Carnwath, and T. B. Chapman. 2009. Implementation of National Fire Plan treatments near the wildland-urban interface in the western United States. *Proceedings of the National Academy of Sciences of the United States of America* 106(26):10706-10711.

Scott, D., and C. Lemieux. 2005. Climate change and protected area policy and planning in Canada. *Forestry Chronicle* 81 (5):696-703.

Seager, R., M. Ting, I. Held, Y. Kushnir, J. Lu, G. Vecchi, H. P. Huang, N. Harnik, A. Leetmaa, N. C. Lau, C. Li, J. Velez, and N. Naik. 2007. Model projections of an imminent transition to a more arid climate in southwestern North America. *Science* 316(5828):1181-1184.

SEI (Stockholm Environment Institute), Å. Persson, R. J.T. Klein, C. Kehler Siebert, A. Atteridge, B. Müller, J. Hoffmaister, M. Lazarus, and T. Takama. 2009. *Adaptation Finance Under a Copenhagen Agreed Outcome*. Stockholm: SEI.

Semenza, J. C., C. H. Rubin, K. H. Falter, J. D. Selanikio, W. D. Flanders, H. L. Howe, and J. L. Wilhelm. 1996. Heat-related deaths during the July 1995 heat wave in Chicago. *New England Journal of Medicine* 335(2):84-90.

Semenza, J. C., J. E. McCullough, W. D. Flanders, M. A. McGeehin, and J. R. Lumpkin. 1999. Excess hospital admissions during the July 1995 heat wave in Chicago. *American Journal of Preventive Medicine* 16(4):269-277.

Sheridan, S. C., and L. S. Kalkstein. 1998. Heat watch warning systems in urban areas. *World Resource Review* 10:375-383.

Slovic, P. 2000. *The Perception of Risk*. London: Earthscan.

Smit, B., O. Pilifosova, I. Burton, B. Challenger, S. Huq, R. J.T. Klein, and G. Yohe. 2001. Adaptation to Climate Change in the Context of Sustainable Development and Equity. In *Climate Change 2001: Working Group II: Impacts, Adaptation and Vulnerability. Contribution of Working Group II to the Third Assessment Report of the Intergovernmental Panel on Climate Change*. New York: Cambridge University Press.

Smith, J. B. 1997. Setting priorities for adapting to climate change. *Global Environmental Change-Human and Policy Dimensions* 7(3):251-264.

Smith, J. B., H.-J. Schellnhuber, M. Monirul Qader Mirza, S. Fankhauser, R. Leemans, L. Erda, L. Ogallo, B. Pittock, R. Richels, C. Rosenzweig, U. Safriel, R. S. J. Tol, J. Weyant, and G. Yohe. 2001. Vulnerability to climate change and reasons for concern: A synthesis. In *Climate Change 2001: Impacts, Adaptation and Vulnerability. Contribution of Working Group II to the Third Assessment Report of the Intergovernmental Panel on Climate Change*. New York: Cambridge University Press.

Smith, J. B., S. H. Schneider, M. Oppenheimer, G. W. Yohe, W. Hare, M. D. Mastrandrea, A. Patwardhan, I. Burton, J. Corfee-Morlot, C. H. D. Magadza, H. M. Fuessel, A. B. Pittock, A. Rahman, A. Suarez, and J. P. van Ypersele. 2009a. Assessing dangerous climate change through an update of the Intergovernmental Panel on Climate Change (IPCC) "reasons for concern." *Proceedings of the National Academy of Sciences of the United States of America* 106(11):4133-4137.

Smith, J. B., J. M. Vogel, and J. E. Cromwell III. 2009b. An architecture for government action on adaptation to climate change. An editorial comment. *Climatic Change* 95:53-61.

Snover, A. K. 2007. Preparing for climate change: A guidebook for local, regional, and state governments. Washington, DC: ICLEI.

So, F. S., I. Hand, and B. D. McDowell. 1986. *The Practice of State and Regional Planning*. Chicago: American Planning Association, International City Management Association.

Solomon, S., G. K. Plattner, R. Knutti, and P. Friedlingstein. 2009. Irreversible climate change due to carbon dioxide emissions. *Proceedings of the National Academy of Sciences of the United States of America* 106(6):1704-1709.

Stainforth, D. A., T. Aina, C. Christensen, M. Collins, N. Faull, D. J. Frame, J. A. Kettleborough, S. Knight, A. Martin, J. M. Murphy, C. Piani, D. Sexton, L. A. Smith, R. A. Spicer, A. J. Thorpe, and M. R. Allen. 2005. Uncertainty in predictions of the climate response to rising levels of greenhouse gases. *Nature* 433:403-406.

Stern, N. H. 2007. *The Economics of Climate Change: The Stern Review.* New York: Cambridge University Press.

Strom, B. A., and P. Z. Fule. 2007. Pre-wildfire fuel treatments affect long-term ponderosa pine forest dynamics. *International Journal of Wildland Fire* 16(1):128-138.

Stutz, B. 2009. Adaptation emerges as key part of any climate change plan. *Yale Environment 360,* 26(May).

Swetnam, T. W., and C. H. Baisan. 1995. *Historical Fire Regime Patterns in the Southwestern United States Since AD 1700.* Tucson, AZ: Laboratory of Tree-Ring Research.

Tainter, J. A. 2003. A framework for sustainability. *World Futures* 59:213-223.

Tarlock, A. D. 2000. How well can international water allocation regimes adapt to global climate change? *Journal of Land Use & Environmental Law,* 15:423-449. Available at *http://www.law.fsu.edu/journals/landuse/Vol153/tarlock.pdf.* Accessed December 30, 2009.

Tebaldi, C., and R. Knutti. 2007. The use of the multi-model ensemble in probabilistic climate projection. *Philosophical Transactions of the Royal Society A* 365:2053-2075.

Theobald, D. M., and W. H. Romme. 2007. Expansion of the US wildland-urban interface. *Landscape and Urban Planning* 83(4):340-354.

Theoharides, K., G. Barnhart, and P. Glick. 2009. Climate Change Adaptation Across the Landscape: A Survey of Federal and State Agencies, Conservation Organizations and Academic Institutions in the United States. The Association of Fish and Wildlife Agencies, Defenders of Wildlife, The Nature Conservancy, and The National Wildlife Federation.

Timilsena, J., et al. 2007. Five hundred years of hydrological drought in the Upper Colorado River Basin. *Journal of the American Water Resources Association* 43(3):798-812.

TNC (The Nature Conservancy). 2009. *The Nature Conservancy on Long Island Addresses Global Warming: Two Projects to Predict, Measure Global Warming, Sea Level Rise Underway (March 12, 2009).* Available at *http://www.nature.org/wherewework/northamerica/states/newyork/press/press3937.html.* Accessed November 2, 2009.

Tobin, G. A., and B. E. Montz. 1997. *Natural Hazards: Explanation and Integration.* New York: Guilford Press.

Tol, R. S. J., and D. Anthoff. 2008. Estimates of the social cost of carbon for the United States, in support of the preparation of U.S. Environmental Protection Agency (EPA) Technical Support Document on Benefits of Reducing GHG Emissions.

Tol, R. S. J., and G. W. Yohe. 2007. The weakest link hypothesis for adaptive capacity: An empirical test. *Global Environmental Change* 17(2):218-227.

Tol, R. S. J., T. E. Downing, O. J. Kuik, and J. B. Smith. 2004. Distributional aspects of climate change impacts. *Global Environmental Change* 14(3):259-272.

Tompkins, E., S. A. Nicholson-Cole, L. A. Hurlston, E. Boyd, G.B. Hodge, J. Clarke, G. Gray, N. Trotz, and L. Varlack, (2005) Surviving climate change in small islands: A guidebook, Norwich: Tyndall Centre for Climate Change Research, University of East Anglia.

Trenberth, K. E., J. T. Houghton, and L. G. Meira Filho. 1995. The climate system: An overview. In *Climate Change 1995: The Science of Climate Change* J. T. Houghton, L. G. Meira Filho, B. A. Callander, N. Harris, A. Kattenberg, and K. Maskell, eds. Contribution of WGI to the Second Assessment Report of the Intergovernmental Panel on Climate Change. Cambridge, UK: Cambridge University Press.

Turner, B. L., R. E. Kasperson, P. A. Matson, J. J. McCarthy, R. W. Corell, L. Christensen, N. Eckley, J. X. Kasperson, A. Luers, M. L. Martello, C. Polsky, A. Pulsipher, and A. Schiller. 2003. A framework for vulnerability analysis in sustainability science. *Proceedings of the National Academy of Sciences of the United States of America* 100(14):8074-8079.

UKCIP (United Kingdom Climate Impacts Program). 2009. United Kingdom Climate Projections (UKCP09). Available at *http://ukclimateprojections.defra.gov.uk/.* Accessed October 15, 2010.

UNDP (UN Development Programme). 2007. *Governance Indicators: A Users' Guide,* 2nd ed. New York: UNDP Oslo Governance Centre.

UNFCCC (UN Framework Convention on Climate Change). 2007. Investment and financial flows relevant to the development of an effective and appropriate international response to *Climate Change 2007*. UNFCCC.

USACE (U.S. Army Corps of Engineers). 2009. *Water Resource Policies and Authorities Incorporating Sea-Level Change Considerations in Civil Works Programs*. Circular No. 1165-2-211. Washington, DC: Department of the Army.

USCAP (U.S. Climate Action Partnership). 2007. *A Call for Action: Consensus Principles and Recommendations from the U.S. Climate Action Partnership: A Business and NGO Partnership*. Available at *http://www.pewclimate.org/docUploads/ USCAP%20Report%20FINAL%20070117.pdf*. Accessed October 15, 2010.

U.S. Conference of Mayors. 2008. Climate change adaptation and vulnerability assessments. The U.S. Conference of Mayors 76th Annual Meeting, June 20-24, Adopted Resolutions.

USGCRP (United States Global Change Research Program). 2001. *Climate Change Impacts on the United States: The Potential Consequences of Climate Variability and Change, Foundation Report*. Washington, DC: USGCRP.

_____. 2009. *Global Climate Change Impacts in the United States: A State of Knowledge Report from the U.S. Global Change Research Program*, T. R. Karl, J. M. Melillo, and T. C. Peterson, eds. New York: Cambridge.

Vance, C., and M. Mehlin. 2009. Fuel costs, circulation taxes, and car market shares implications for climate policy. *Transportation Research Record* (2134):31-36.

Walters, C., and R. Ahrens. 2009. Oceans and estuaries: Managing the commons. In *Principles of Ecosystems Stewardship: Resilience-based natural resource management in a changing world*. F. S. Chapin, ed. New York: Springer.

Wang, Y. Q., L. R. Leung, J. L. McGregor, D. K. Lee, W. C. Wang, Y. H. Ding, and F. Kimura. 2004. Regional climate modeling: Progress, challenges, and prospects. *Journal of the Meteorological Society of Japan* 82(6):1599-1628.

Warner, J. 2009. Senate Foreign Relations Committee. *Testimony by Senator John Warner (Retired)*. July 21.

WCGA (West Coast Governors Association), 2008. *West Coast Governors' Agreement on Ocean Health—Action Plan*. Christine Gregoire (Washington), Theodore Kulongoski (Oregon), Arnold Schwarzenegger (California). May 2008. *http:// westcoastoceans.gov/docs/WCGA_ActionPlan_lowest-resolution.pdf*.

Webster, M. D., C. Forest, J. M. Reilly, M. Babiker, D. Kicklighter, M. Mayer, R. Prinn, M. C. Sarofim, A. Sokolov, P. Stone, and C. Wang. 2003. Uncertainty analysis of climate change and policy response. *Climatic Change* 61:295-320.

Weisskopf, M. G., H. A. Anderson, S. Foldy, L. P. Hanrahan, K. Blair, T. J. Torok, and P. D. Rumm. 2002. Heat wave morbidity and mortality, Milwaukee, Wis, 1999 vs 1995: An improved response? *American Journal of Public Health* 92(5):830-833.

West, J. M., S. H. Julius, P. Kareiva, C. Enquist, J. J. Lawler, B. Petersen, A. E. Johnson, and M. R. Shaw. 2009. U.S. natural resources and climate change: Concepts and approaches for management adaptation. *Environmental Management* 1-21.

Westerling, A. L., H. G. Hidalgo, D. R. Cayan, and T. W. Swetnam. 2006. Warming and earlier spring increase western US forest wildfire activity. *Science* 313 (5789):940-943.

Whitman, S., G. Good, E. R. Donoghue, N. Benbow, W. Y. Shou, and S. X. Mou. 1997. Mortality in Chicago attributed to the July 1995 heat wave. *American Journal of Public Health* 87(9):1515-1518.

Wigley, T. M. L., and S. C. B. Raper. 2001. Interpretation of high projections for global-mean warming. *Science* 293:451-454.

Wigley, T. M. L., R. G. Richels, and J. A. Edmonds. 1996. Economic and environmental choices in the stabilisation of atmospheric CO_2 concentrations. *Nature* 379:240-243.

Wilbanks, T., and J. Sathaye. 2007. Toward an integrated analysis of mitigation and adaptation: Some preliminary findings. Challenges in Integrating Mitigation and Adaptation as Responses to Climate Change, Special Issue. *Mitigation and Adaptation Strategies for Global Change* 12(5):713-725.

Wolf, A. 2007. Shared waters: Conflict and cooperation. *Annual Review of Environment and Resources* 32: 241-269.

Wolf, A., J. Natharius, J. Danielson, B. Ward, and J. Pender. 1999. International river basins of the world. *International Journal of Water Resource Development* 15(4):387-427.

Wolf, A., S. Yoffe, and M. Giordano. 2003. International waters: Identifying basins at risk. *Water Policy* 5(1):31-62.

Woolsey, R. J. 2009. *Threats to National Security: Developing the Framework for a National Response to Climate Change*. Presentation at the Summit on America's Climate Choices, March 30, 2009. National Academy of Sciences, Washington, DC.

World Bank. 2009. *The Costs to Developing Countries of Adapting to Climate Change: New Methods and Estimates Consultation Draft*. The Global Report of the Economics of Adaptation to Climate Change Study. Washington, DC: World Bank.

———. 2010. *World Development Report 2010: Development and Climate Change*. Washington, DC: World Bank.

WWF (World Wildlife Federation). 2003. *Buying Time: A User's Manual for Building Resistance and Resilience to Climate Change in Natural Systems*, L. J. Hansen, J. L. Biringer, and J. R. Hoffman, eds. Washington, DC: WWF.

Yohe, G. 2009a. *Addressing Climate Change Through a Risk Management Lens—An Overview of Analytic Approaches for Climate Change Based on a Deconstruction of Synthetic Conclusions of the Fourth Assessment Report of the Intergovernmental Panel on Climate Change.* Washington, DC: Pew Center on Global Climate Change.

———. 2009b. Toward an integrated framework derived from a risk-management approach to climate change. *Climatic Change* 95(3-4):325-339.

———. 2010. "Reasons for Concern" (about Climate Change) in the United States. *Climatic Change* 99:295-302.

Yohe, G., and D. Tirpak. 2008. Summary report: OECD global forum on sustainable development: The economic benefits of climate change policies. *Integrated Assessment Journal* 8:1-17.

Young, M., and J. McColl. 2008. *A Future-Proofed Basin: A New Water Management Regime for the Murray-Darling Basin.* Adelaide, Australia: University of Adelaide.

Young, M., and J. McColl. 2009. *Droplet 17: Water Security: Should Urban Water Use, Like Rural Water Use, Be Capped? May 24, 2009.* Water Droplets: The University of Adelaide.

Zhang, X. B., F. W. Zwiers, G. C. Hegerl, F. H. Lambert, N. P. Gillett, S. Solomon, P. A. Stott, and T. Nozawa. 2007. Detection of human influence on twentieth-century precipitation trends. *Nature* 448(7152):461-464.

America's Climate Choices: Membership Lists

COMMITTEE ON AMERICA'S CLIMATE CHOICES

ALBERT CARNESALE (Chair), University of California, Los Angeles
WILLIAM CHAMEIDES (Vice Chair), Duke University, Durham, North Carolina
DONALD F. BOESCH, University of Maryland Center for Environmental Science, Cambridge
MARILYN A. BROWN, Georgia Institute of Technology, Atlanta
JONATHAN CANNON, University of Virginia, Charlottesville
THOMAS DIETZ, Michigan State University, East Lansing
GEORGE C. EADS, Charles River Associates, Washington, D.C.
ROBERT W. FRI, Resources for the Future, Washington, D.C.
JAMES E. GERINGER, Environmental Systems Research Institute, Cheyenne, Wyoming
DENNIS L. HARTMANN, University of Washington, Seattle
CHARLES O. HOLLIDAY, JR., DuPont, Wilmington, Delaware
KATHARINE L. JACOBS,* Arizona Water Institute, Tucson
THOMAS KARL,* National Oceanic and Atmospheric Administration, Asheville, North Carolina
DIANA M. LIVERMAN, University of Arizona, Tuscon and University of Oxford, United Kingdom
PAMELA A. MATSON, Stanford University, California
PETER H. RAVEN, Missouri Botanical Garden, St. Louis
RICHARD SCHMALENSEE, Massachusetts Institute of Technology, Cambridge
PHILIP R. SHARP, Resources for the Future, Washington, D.C.
PEGGY M. SHEPARD, WE ACT for Environmental Justice, New York, New York
ROBERT H. SOCOLOW, Princeton University, New Jersey
SUSAN SOLOMON, National Oceanic and Atmospheric Administration, Boulder, Colorado
BJORN STIGSON, World Business Council for Sustainable Development, Geneva, Switzerland

THOMAS J. WILBANKS, Oak Ridge National Laboratory, Tennessee
PETER ZANDAN, Public Strategies, Inc., Austin, Texas

PANEL ON LIMITING THE MAGNITUDE OF FUTURE CLIMATE CHANGE

ROBERT W. FRI (Chair), Resources for the Future, Washington, D.C.
MARILYN A. BROWN (Vice Chair), Georgia Institute of Technology, Atlanta
DOUG ARENT, National Renewable Energy Laboratory, Golden, Colorado
ANN CARLSON, University of California, Los Angeles
MAJORA CARTER, Majora Carter Group, LLC, Bronx, New York
LEON CLARKE, Joint Global Change Research Institute (Pacific Northwest National Laboratory/University of Maryland), College Park, Maryland
FRANCISCO DE LA CHESNAYE, Electric Power Research Institute, Washington, D.C.
GEORGE C. EADS, Charles River Associates, Washington, D.C.
GENEVIEVE GIULIANO, University of Southern California, Los Angeles
ANDREW J. HOFFMAN, University of Michigan, Ann Arbor
ROBERT O. KEOHANE, Princeton University, New Jersey
LOREN LUTZENHISER, Portland State University, Oregon
BRUCE MCCARL, Texas A&M University, College Station
MACK MCFARLAND, DuPont, Wilmington, Delaware
MARY D. NICHOLS, California Air Resources Board, Sacramento
EDWARD S. RUBIN, Carnegie Mellon University, Pittsburgh, Pennsylvania
THOMAS H. TIETENBERG, Colby College (retired), Waterville, Maine
JAMES A. TRAINHAM, RTI International, Research Triangle Park, North Carolina

PANEL ON ADAPTING TO THE IMPACTS OF CLIMATE CHANGE

KATHARINE L. JACOBS* (Chair, through January 3, 2010), University of Arizona, Tucson
THOMAS J. WILBANKS (Chair), Oak Ridge National Laboratory, Tennessee
BRUCE P. BAUGHMAN, IEM, Inc., Alabaster, Alabama
ROBERT BEACHY,* Donald Danforth Plant Sciences Center, Saint Louis, Missouri
GEORGES C. BENJAMIN, American Public Health Association, Washington, D.C.
JAMES L. BUIZER, Arizona State University, Tempe
F. STUART CHAPIN III, University of Alaska, Fairbanks
W. PETER CHERRY, Science Applications International Corporation, Ann Arbor, Michigan
BRAXTON DAVIS, South Carolina Department of Health and Environmental Control, Charleston
KRISTIE L. EBI, IPCC Technical Support Unit WGII, Stanford, California

JEREMY HARRIS, Sustainable Cities Institute, Honolulu, Hawaii
ROBERT W. KATES, Independent Scholar, Bangor, Maine
HOWARD C. KUNREUTHER, University of Pennsylvania Wharton School of Business, Philadelphia
LINDA O. MEARNS, National Center for Atmospheric Research, Boulder
PHILIP MOTE, Oregon State University, Corvallis
ANDREW A. ROSENBERG, Conservation International, Arlington, Virginia
HENRY G. SCHWARTZ, JR., Jacobs Civil (retired), Saint Louis, Missouri
JOEL B. SMITH, Stratus Consulting, Inc., Boulder, Colorado
GARY W. YOHE, Wesleyan University, Middletown, Connecticut

PANEL ON ADVANCING THE SCIENCE OF CLIMATE CHANGE

PAMELA A. MATSON (Chair), Stanford University, California
THOMAS DIETZ (Vice Chair), Michigan State University, East Lansing
WALEED ABDALATI, University of Colorado at Boulder, Colorado
ANTONIO J. BUSALACCHI, JR., University of Maryland, College Park
KEN CALDEIRA, Carnegie Institution of Washington, Stanford, California
ROBERT W. CORELL, H. John Heinz III Center for Science, Economics and the Environment, Washington, D.C.
RUTH S. DEFRIES, Columbia University, New York, New York
INEZ Y. FUNG, University of California, Berkeley
STEVEN GAINES, University of California, Santa Barbara
GEORGE M. HORNBERGER, Vanderbilt University, Nashville, Tennessee
MARIA CARMEN LEMOS, University of Michigan, Ann Arbor
SUSANNE C. MOSER, Susanne Moser Research & Consulting, Santa Cruz, California
RICHARD H. MOSS, Joint Global Change Research Institute (Pacific Northwest National Laboratory/University of Maryland), College Park, Maryland
EDWARD A. PARSON, University of Michigan, Ann Arbor
A. R. RAVISHANKARA, National Oceanic and Atmospheric Administration, Boulder, Colorado
RAYMOND W. SCHMITT, Woods Hole Oceanographic Institution, Massachusetts
B. L. TURNER II, Arizona State University, Tempe
WARREN M. WASHINGTON, National Center for Atmospheric Research, Boulder, Colorado
JOHN P. WEYANT, Stanford University, California
DAVID A. WHELAN, The Boeing Company, Seal Beach, California

PANEL ON INFORMING EFFECTIVE DECISIONS AND
ACTIONS RELATED TO CLIMATE CHANGE

DIANA LIVERMAN (Co-chair), University of Arizona, Tucson

PETER RAVEN (Co-chair), Missouri Botanical Garden, Saint Louis

DANIEL BARSTOW, Challenger Center for Space Science Education, Alexandria, Virginia

ROSINA M. BIERBAUM, University of Michigan, Ann Arbor

DANIEL W. BROMLEY, University of Wisconsin-Madison

ANTHONY LEISEROWITZ, Yale University

ROBERT J. LEMPERT, The RAND Corporation, Santa Monica, CA

JIM LOPEZ,* King County, Washington

EDWARD L. MILES, University of Washington, Seattle

BERRIEN MOORE III, Climate Central, Princeton, New Jersey

MARK D. NEWTON, Dell, Inc., Round Rock, Texas

VENKATACHALAM RAMASWAMY, National Oceanic and Atmospheric Administration, Princeton, New Jersey

RICHARD RICHELS, Electric Power Research Institute, Inc., Washington, D.C.

DOUGLAS P. SCOTT, Illinois Environmental Protection Agency, Springfield

KATHLEEN J. TIERNEY, University of Colorado at Boulder

CHRIS WALKER, The Carbon Trust LLC, New York, New York

SHARI T. WILSON, Maryland Department of the Environment, Baltimore

Asterisks (*) denote members who resigned during the study process

Panel on Adapting to the Impacts of Climate Change: Statement of Task

The Panel on Adapting to the Impacts of Climate Change will describe, analyze, and assess actions and strategies to reduce vulnerability, increase adaptive capacity, improve resiliency, and promote successful adaptation to climate change in different regions, sectors, systems, and populations. The panel will draw on existing reports and assessments and use case studies to identify lessons learned from past experiences, promising current approaches, and potential new directions. The issues and examples considered by the panel should be drawn from a variety of regions and sectors, focusing on domestic actions but also considering international dimensions, and should cover a range of temporal and spatial scales.

The panel will be challenged to produce a report that is broad and authoritative, yet concise and useful to decision makers. The costs, benefits, limitations, tradeoffs, and uncertainties associated with different options and strategies should be assessed qualitatively and, to the extent practicable, quantitatively, using the scenarios of future climate change and vulnerability provided by the Climate Change Study Committee. The panel should also provide policy-relevant (but not policy-prescriptive) input to the committee on the following overarching questions:

- What short-term actions can be taken to adapt effectively to climate change?
- What promising long-term strategies, investments, and opportunities could be pursued to adapt to climate change?
- What are the major scientific and technological advances (e.g., new observations, improved models, research priorities, etc.) needed to promote effective adaptation to climate change?
- What are the major impediments (e.g., practical, institutional, economic, ethical, intergenerational, etc.) to effective adaptation to climate change, and what can be done to overcome these impediments?
- What can be done to adapt to climate change at different levels (e.g., local, state, regional, national, and in collaboration with the international community) and in different sectors (e.g., nongovernmental organizations, the business community, the research and academic communities, individuals and households, etc.)?

Panel on Adapting to the Impacts of Climate Change: Biographical Sketches

Katharine L. Jacobs* (*Chair through January 3, 2010*) is a Professor in the University of Arizona Department of Soil, Water and Environmental Science and an Associate Director of the National Science Foundation (NSF) Center for Sustainability of Arid Region Hydrology and Riparian Areas at the University of Arizona. She is affiliated with the recently established Institute of the Environment, working on climate adaptation and water management issues. For the past 3 years, Jacobs was the Executive Director of the Arizona Water Institute, a consortium of the three state universities focused on water-related research, education, and technology transfer in support of water supply sustainability. She has more than 20 years of experience as a water manager for the State of Arizona Department of Water Resources, including 14 years as director of the Tucson Active Management Area. Her research interests include water policy, connecting science and decision making, stakeholder engagement, use of climate information for water management applications, climate-change adaptation, and drought planning. Ms. Jacobs earned her M.L.A. in environmental planning from the University of California, Berkeley. She was the author of the water sector chapter for the first National Assessment of the Impacts of Climate Change, and a convening lead author of the Climate Change Science Program's report *Decision-Support Experiments for Water Resources*. She has served on eight National Research Council panels. She recently testified in the U.S. Senate, providing recommendations on the design of the National Climate Service.

Thomas J. Wilbanks (*Chair*) is a Corporate Research Fellow at the Oak Ridge National Laboratory and leads the Laboratory's Global Change and Developing Country Programs. A past president of the Association of American Geographers, he conducts research on such issues as sustainable development, energy and environmental technology and policy, responses to global climate change, and the role of geographical scale in all of these regards. Wilbanks has won the James R. Anderson Medal of

---

Asterisks (*) denote members who resigned during the study process.

Honor in Applied Geography; has been awarded Honors by the Association of American Geographers, geography's highest honor; was named Distinguished Geography Educator of the year in 1993 by the National Geographic Society; and is a fellow of the American Association for the Advancement of Science (AAAS). Co-edited recent books include *Global Change and Local Places* (2003), *Geographical Dimensions of Terrorism* (2003), and *Bridging Scales and Knowledge Systems: Linking Global Science and Local Knowledge* (2006). Wilbanks is Chair of the National Research Council's (NRC's) Committee on Human Dimensions of Global Change and a member of a number of other National Academy of Sciences (NAS)/NRC boards and panels. In recent years, he has been coordinating lead author for the Intergovernmental Panel on Climate Change (IPCC) Fourth Assessment Report, Working Group II, Chapter 7 (Industry, Settlement, and Society); coordinating lead author for the Climate Change Science Program's Synthesis and Assessment Product (SAP) 4.5 (*Effects of Climate Change on Energy Production and Use in the United States*); and lead author for one of three sections (Effects of Global Change on Human Settlements) of SAP 4.6 (*Effects of Global Change on Human Health and Welfare and Human Systems*). Wilbanks received his B.A. degree in social sciences from Trinity University in 1960 and his M.A. and Ph.D. degrees in geography from Syracuse University in 1967 and 1969.

Bruce Baughman, for more than three decades, has served in key federal and state emergency management positions for some of the largest natural and man-made disasters ever to hit the United States and its territories, including 13 hurricanes, the Oklahoma City bombing, and the 9/11 terrorist attacks. He has testified before Congress on emergency management issues more than 25 times. As director of the Alabama Emergency Management Agency, he led the state's response to three hurricanes—Ivan in 2004 and Dennis and Katrina in 2005—and a deadly series of tornadoes in March of 2007. Prior to his appointment by the governor of Alabama, Mr. Baughman held several key positions at the Federal Emergency Management Agency (FEMA), including Director of the Office of National Preparedness and Director of Operations. While at FEMA, he directed response operations for more than 110 presidential disaster and emergency declarations, including hurricanes, earthquakes, bombings, and flooding. He retired from the Department of Homeland Security as one of FEMA's top senior executives in 2003. Mr. Baughman is the recipient of numerous national awards, including FEMA's Distinguished Service Award, four FEMA Meritorious Service Awards, the National Hurricane Conference's Distinguished Service Award and the Neil Frank Award, and the President's Council on Year 2000 Gold Medal. He is a past president of the National Emergency Management Association. Mr. Baughman is currently Senior Consultant for Emergency Management and Homeland Security to Innovative Emergency Management, Inc.

Roger N. Beachy* is the founding president of the not-for-profit Donald Danforth Plant Science Center in St. Louis, Missouri, a position he has held since January 1999. In this role, Dr. Beachy has been responsible for developing and implementing the Danforth Center's strategic direction, recruiting its staff, and formulating its research programs. Dr. Beachy, a member of the NAS, is internationally known for his ground-breaking research on developing virus-resistant plants through biotechnology. He was a member of the Biology Department at Washington University in St. Louis from 1978 to 1991, where he was Professor and Director of the Center for Plant Science and Biotechnology. His work at Washington University, in collaboration with Monsanto Company, led to the development of the world's first genetically modified food crop, a variety of tomato that was modified for resistance to virus disease. His technique to produce virus resistance in tomatoes has been replicated by researchers around the world to produce many types of plants with resistance to a number of different virus diseases. Research under Dr. Beachy's direction has led to a number of patent applications. He has edited or contributed to 50 book articles, and his work has produced more than 220 journal publications.

Georges C. Benjamin, M.D., F.A.C.P., F.A.C.E.P. (E), is well known in the world of public health as a leader, practitioner, and administrator. Benjamin has been the executive director of the American Public Health Association, the nation's oldest and largest organization of public health professionals, since December 2002. He came to that post from his position as secretary of the Maryland Department of Health and Mental Hygiene, where he played a key role in the expansion and improvement of the Maryland Medicaid program. Benjamin became secretary of the Maryland health department in April 1999, following 4 years as its deputy secretary for public health services. Benjamin, of Gaithersburg, Maryland, is a graduate of the Illinois Institute of Technology and the University of Illinois College of Medicine. He is board certified in internal medicine and a fellow of the American College of Physicians; he is also a Fellow Emeritus of the American College of Emergency Physicians.

James L. Buizer is Science Policy Advisor to the President at Arizona State University (ASU) and Director for Strategic Institutional Transformation in the Office of the President. He also serves as Director of the University Center for Integrated Solutions to Climate Challenges and Professor of Practice in Climate Adaptation Policy & Institutional Design in the School of Geographical Sciences and Urban Planning. Mr. Buizer advances ASU by providing leadership and strategic advice on a broad range of topics. Upon arriving at ASU in September 2003 until July 2007, he served as founding Executive Director of the Office of Sustainability Initiatives in the Office of the President, where he led the conceptualization, design, and initiation of the university-wide Global Institute of Sustainability and its School of Sustainability, launched in fall 2006.

He serves on numerous leadership boards and advisory councils across the university. In his personal capacity he serves on the Board of Directors at the National Council for Science and the Environment; on the Board of Directors of Second Nature, Inc.; on the Board of Trustees of the Tesseract School in Paradise Valley, Arizona; on the Advisory Committee of the American College and University Presidents Climate Commitment; and as Strategic Advisor to Pegasus Capital Advisors, L.P. Prior to this, he served as Director of the Climate and Societal Interactions Office at the National Oceanic and Atmospheric Administration (NOAA) in Washington, D.C. In this capacity Jim coordinated the U.S. government technical review of the 2000 *Assessment Report of the Working Group on Impacts of Climate Change* of the IPCC. Mr. Buizer has presented and published extensively on institutionalizing the science-to-action interface. He received his degrees in oceanography, marine resource economics, and science policy from the University of Washington, Seattle.

F. Stuart Chapin III focuses his research on ecosystem ecology and on the resilience of social-ecological systems. His ecological research addresses the consequences of plant traits for ecosystem and global processes, particularly vegetation effects on nutrient cycling, fire regime, and biodiversity. He also studies vegetation-mediated feedbacks to high-latitude climate warming, as mediated by changes in water and energy exchange. Dr. Chapin's research on social-ecological systems emphasizes the resilience of northern regions to recent changes in climate and fire regime. This research entails studies of human and climatic effects on fire regime; the resulting effects on ecosystem services, wages, and cultural integrity; and the effects of local opinions about fire and national fire policy on the fire policies developed and implemented at regional scales. Most of his current research focuses on Alaska and eastern Siberia. Dr. Chapin has served on numerous NRC committees and is a member of NAS.

W. Peter Cherry is Chief Analyst at Science Applications International Corporation, where his research interests include the design, development, and test and evaluation of large-scale systems with emphasis on network centricity. A member of the National Academy of Engineering (NAE), he has focused on the development and application of operations research in the national security domain, primarily in the field of land combat. He contributed to the development and fielding of most of the major systems currently employed by the Army, ranging from the Patriot Missile System to the Apache helicopter, as well as the command control and intelligence systems currently in use. In addition, he contributed to the creation of the Army's Manpower Personnel and Human Factors and Training Program and to the Army's Embedded Training Initiative. Dr. Cherry is a member of the Board on Army Science and Technology, served on the Army Science Board, and for the past 10 years has participated in independent reviews of the Army's Science and Technology programs.

Braxton Davis is the Director of the Policy and Planning Division for South Carolina's Coastal Zone Management Program, where he leads long-term state policy initiatives focused on shoreline change, ocean resources, and coastal trends analysis. For the past 6 years, he has also worked with NOAA and the Coastal States Organization (CSO) on several national studies related to coastal and ocean policy issues. As a delegate for the state of South Carolina, he currently serves as vice chair of CSO, and he previously served as chair of a Climate Change Work Group that brings coastal states' perspectives on climate change research and policy needs to the federal government. He has provided congressional testimony on climate change issues and continues to serve on several interagency committees to improve federal, state, and local coordination on coastal issues and climate change. Dr. Davis earned a B.S. degree in environmental sciences from the University of Virginia, an M.S. degree in biological sciences from Florida International University, and a Ph.D. in marine affairs from the University of Rhode Island.

Kristie L. Ebi is Executive Director of the Technical Support Unit for Working Group II (Impacts, Adaptation, and Vulnerability) of the IPCC. Prior to this position, she was an independent consultant researching the impacts of and adaptation to climate change for extreme events, thermal stress, foodborne safety and security, and vector-borne diseases. She has worked with the World Health Organization, the United Nations Development Programme, the U.S. Agency for International Development (USAID), and others on implementing adaptation measures in low-income countries. She facilitated adaptation assessments for the health sector for the states of Maryland and Alaska. She was a lead author on the "Human Health" chapter of the IPCC Fourth Assessment Report, and the "Human Health" chapter for the U.S. Synthesis and Assessment Product "Analyses of the Effects of Global Change on Human Health and Welfare and Human Systems." She has edited four books on aspects of climate change and has more than 80 publications. Dr. Ebi's scientific training includes an M.S. in toxicology and a Ph.D. and a master's of public health in epidemiology, and 2 years of postgraduate research at the London School of Hygiene and Tropical Medicine.

Jeremy Harris served for more than 10 years as the Mayor of the City and County of Honolulu, Hawaii, the 12th largest city in the United States. He retired in January of 2005. Prior to becoming mayor, he was Honolulu's longest serving managing director, a position he held for almost 9 years. Under his leadership, Honolulu received the Gold Award as the most livable large city in the world. Mayor Harris is the only individual to receive the award of Public Administrator of the Year for two consecutive years from the American Association of Public Administrators in Hawaii. He has served on the board of directors of the national American Institute of Architects, as Irving Distinguished Professor at Ball State University, and as visiting senior faculty at the Royal

Institute of Technology in Stockholm, Sweden. He holds an M.S. degree in population and environmental biology, specializing in urban ecosystems, from the University of California, Irvine, and is the author of the book *The Renaissance of Honolulu, the Sustainable Rebirth of an American City*.

Robert W. Kates is a Senior Research Associate at Harvard University, Presidential Professor of Sustainability Science at the University of Maine, and University Professor (Emeritus) at Brown University. Trained as a geographer, he has led interdisciplinary programs addressing hazards, climate, and adaptation at the University of Dar as Salaam in Tanzania, Clark University, and the World Hunger Program at Brown University. He has participated in all four IPCC Assessments, in the NRC Committee on Global Change, and in State of Maine climate advisory groups. He has co-authored or edited foundational studies on natural hazards, on climate impact assessment, and on global change in local places. His most recent research is on reconstruction following Hurricane Katrina and his current research is on enhancing community resilience to multiple hazards. Dr. Kates is a member of the NAS.

Howard C. Kunreuther is the Cecilia Yen Koo Professor of Decision Sciences and Public Policy at the Wharton School, Co-Director of the Wharton Risk Management and Decision Processes Center. He has a long-standing interest in ways that society can better manage low-probability, high-consequence events related to technological and natural hazards and has published widely in these areas. Dr. Kunreuther is a Fellow of the AAAS, a member of the NAS Panel on Adaptation Strategies for Climate Change, and Distinguished Fellow of the Society for Risk Analysis, receiving the Society's Distinguished Achievement Award in 2001. He co-chaired the World Economic Forum's Global Agenda Council on "Innovation and Leadership in Reducing Risks from Natural Disasters" and is a member of the Organisation for Economic Co-operation and Development's High Level Advisory Board on Financial Management of Large-Scale Catastrophes. His most recent books are *At War with the Weather* (with Erwann Michel-Kerjan, July 2009, MIT Press) and *Learning from Catastrophes: Strategies for Reaction and Response* (with Michael Useem, December 2009, Wharton School Publishing).

Linda O. Mearns is a Senior Scientist in the Institute for the Study of Society and Environment (ISSE) at the National Center for Atmospheric Research, Boulder, Colorado. She served as Director of ISSE for 3 years ending in April 2008. She holds a Ph.D. in geography/climatology from the University of California, Los Angeles. She has performed research and published mainly in the areas of climate change scenario formation, quantifying uncertainties, and climate change impacts on agro-ecosystems. She has particularly worked extensively with regional climate models. She has most recently published papers on the effect of uncertainty in climate change scenarios on agricultural and economic impacts of climate change, and quantifying uncertainty of regional climate change. She has been an author in the IPCC Climate Change 1995, 2001, and 2007 Assessments regarding climate variability, impacts of climate change on agriculture, regional projections of climate change, climate scenarios, and uncertainty in future projections of climate change. For the 2007 Report(s) she was lead author for the chapter on regional projections of climate change in Working Group 1 and for the chapter on new assessment methods in Working Group 2. She is also an author on two Synthesis Products of the U.S. Climate Change Science Program. She leads the multiagency-supported North American Regional Climate Change Assessment Program, which is providing multiple high-resolution climate change scenarios for the North American impacts community. She is a member of the NRC Climate Research Committee and Human Dimensions of Global Change Committee. She was made a Fellow of the American Meteorological Society in January 2006.

Philip Mote serves as Director of the Oregon Climate Change Research Institute and Oregon Climate Services at Oregon State University and is a full professor in the College of Oceanic and Atmospheric Sciences. Until July 2009 he also worked at University of Washington (UW) as a research scientist with the Climate Impacts Group, where since 1998 he had built the group's public profile through hundreds of public speaking events, over a thousand media interviews, deep engagement with the region's stakeholders, and groundbreaking research in the impacts of climate change on the West's mountain snow and on wildfire. He has published over 70 scientific articles and edited a book on climate modeling. He served as state climatologist for Washington and, as Director of Oregon Climate Services, serves in a similar role there. He was a lead author of the IPCC Fourth Assessment Report; the IPCC was awarded the Nobel Peace Prize in 2007. In 2008 he received the UW Distinguished Staff Award and was named one of the region's 25 most influential people by *Seattle Magazine*. He earned a Ph.D. in atmospheric sciences from UW and a B.A. in physics from Harvard.

Andrew A. Rosenberg is Senior Vice President for Science and Knowledge for Conservation International and Professor in the Institute for the Study of Earth, Oceans, and Space at the University of New Hampshire where, prior to April 2004, he was Dean of the College of Life Sciences and Agriculture. From 2001 to 2004, he was a member of the U.S. Commission on Ocean Policy and continues to work with the U.S. Joint Ocean Commissions Initiative. Dr. Rosenberg was the Deputy Director of NOAA's National Marine Fisheries Service (NMFS) from 1998 to 2000, the senior career position in the agency, and prior to that he was the NMFS Northeast Regional Administrator. Dr. Rosenberg's scientific work is in the field of population dynamics, resource assessment, and resource management policy. He holds a B.S. in fisheries biology from the University of Massachusetts, an M.S. in oceanography from Oregon State University, and a Ph.D. in biology from Dalhousie University.

Henry G. Schwartz, Jr., is an independent consultant. He is a nationally recognized civil and environmental engineering leader who spent most of his career with Sverdrup Civil, Inc. (now Jacobs Civil, Inc.), which he joined as a registered professional engineer in 1966. In 1993, Schwartz was named president and chairman, directing the transportation, public works, and environmental activities of Sverdrup/Jacobs Civil, Inc., before he retired in 2003. Dr. Schwartz's projects included multibillion-dollar water and wastewater treatment systems for the cities of San Diego, San Francisco, and Detroit as well as large civil-infrastructure projects, such as highways, bridges, dams, and railroads. Dr. Schwartz is a Director of the Berger Group and was a Senior Professor of Engineering Management at Washington University in St. Louis from 2003 to 2007. He has served on the advisory boards for Carnegie Mellon University, Washington University, and the University of Texas, and he is President Emeritus of the Academy of Science of St. Louis. He is Founding Chairman of the Water Environment Research Foundation and served as President of the Water Environment Federation. Dr. Schwartz is past president of the American Society of Civil Engineers. He was elected to NAE in 1997 (Section 4: civil engineering) and has served on several NRC study committees, including service as chair of the Committee on Climate Change and U.S. Transportation, and is on the Executive Committee of the Transportation Research Board. Currently, he is a member of the Unified Synthesis Product Development Committee of the U.S. Climate Change Science Program. Dr. Schwartz received a Ph.D. from the California Institute of Technology and M.S. and B.S. degrees from Washington University; he also attended Princeton University and Columbia University's Business Program.

Joel B. Smith has been analyzing climate change impacts and adaptation issues since 1987. He was a coordinating lead author for the synthesis chapter on climate change impacts for the Third Assessment Report of the Intergovernmental Panel on Climate Change and a lead author on a similar chapter in the Fourth Assessment, a lead author for the U.S. National Assessment on climate change impacts, technical coordinator on vulnerability and adaptation for the U.S. Country Studies Program, and coordinator of the Pew Center on Global Climate Change series on environment. He has provided technical advice, guidance, and training on assessing climate change impacts and adaptation to people around the world and for clients such as the United Nations; the World Bank; the U.S. Environmental Protection Agency (EPA); USAID; the states of California, Florida, and Alaska; and for municipalities such as Denver, San Francisco, Phoenix, and Boulder. Mr. Smith worked for the EPA from 1984 to 1992, where he was the deputy director of Climate Change Division. He is a co-editor of EPA's Report to Congress: The Potential Effects of Global Climate Change on the United States (1989);

As Climate Changes: International Impacts and Implications (Cambridge University Press, 1995); *Adaptation to Climate Change: Assessments and Issues* (Springer-Verlag, 1996); and *Climate Change, Adaptive Capacity, and Development* (Imperial College Press, 2003). He joined Hagler Bailly in 1992 and Stratus Consulting in 1998. He has published more than 35 articles and chapters on climate change impacts and adaptation in peer-reviewed journals and books. Besides working on climate change issues at EPA, he also was a special assistant to the assistant administrator for the Office of Policy, Planning, and Evaluation. Mr. Smith was a presidential management intern in the Office of the Secretary of Defense from 1982 to 1984. He has also worked in the U.S. Department of Energy and USAID. Joel Smith received an M.P.P., B.A. in political science.

Gary W. Yohe is the Woodhouse/Sysco Professor of Economics at Wesleyan University. He was educated at the University of Pennsylvania and received his Ph.D. in economics from Yale University in 1975. He is the author of more than 100 scholarly articles, several books, and many contributions to media coverage of climate issues. Most of his work has focused attention on the mitigation and adaptation/impact sides of the climate issue. Dr. Yohe served as Convening Lead Author for one chapter in the Response Options Technical Volume of the Millennium Ecosystem Assessment; it focused on uncertainty and the evaluation of response options. Recognizing the enormous uncertainty with which we view the future evolution of the climate and socioeconomic systems led him to call for a risk-management approach to climate policy—an approach that was ultimately adopted in the Synthesis Report of the Fourth Assessment Report of the IPCC in 2007. He has been a senior member of the IPCC since the mid-1990s, serving as a lead author for four different chapters in the Third Assessment Report and as convening lead author for the last chapter of the contribution of Working Group II to the Fourth Assessment Report (AR4). He also worked with the core writing team to prepare the overall Synthesis Report for the entire AR4. Dr. Yohe also served as one of five editors of Avoiding Dangerous Climate Change, and he has testified before the Senate Foreign Relations Committee on the "Hidden (climate change) Cost of Oil" on March 30, 2006; the Senate Energy Committee on the Stern Review on February 14, 2007; and the Senate Banking Committee on Material Risk from Climate Change and Climate Policy on October 31, 2007. He sits currently on the New York Panel on Climate Change, the Committee on the Human Dimensions of Global Change for the NRC, and the Committee on Stabilization Targets for Atmospheric Greenhouse Gas Concentrations, also for the NRC.

Explanation of the Rationale for Reasons of Concern

This appendix provides the detailed assumptions and explanation for Figure 2.9 in Chapter 2. For the United States, assessments of droughts, wildfires, heat waves, extreme precipitation events, and pervasive outbreaks of pests (reported in both IPCC [2007b] and USGCRP [2009]) support the first column for the risks of weather extreme events. For each of these extreme events, both assessments report an increase in frequency over the past few decades that can be attributed to warming and precipitation trends. This column therefore begins yellow and turns orange at around 3.6°F (2°C) to reflect the growing range of pine-beetle destruction across the western states and into southern Alaska and the associated heightened threat of extraordinarily dangerous wildfires (Westerling et al., 2006), as well as the anticipated acceleration of other effects. These high-risk impacts will be encountered roughly at the middle of the midcentury temperature range for the A1b SRES scenario upon which these projections were based.

 The second column in the graphic relates to climate-borne risks to unique and threatened (human and natural) systems across the United States, which were reported by Fischlin et al. (2007) and the National Science and Technology Council (NSTC, 2008). The links with climate are complex and diverse, but it is possible to detect a common thread. Rosenzweig et al. (2008) showed a concentration of statistically significant observed impacts in the West, providing evidence that this reason for concern should begin yellow or perhaps orange. Threats facing Arctic communities from eroding coasts are the result of coastal storms superimposed upon rising seas. Corals face death and eventual collapse from bleaching episodes that are the result of short periods of unusually high ocean temperatures in combination with other factors such as eutrophication and ocean acidification. Coastal wetlands, and the protection that they provide, can be completely destroyed by the storm surges of high-intensity storms. Even without expanding this list of examples, it is clear that a wide range of climate-related risks to unique and threatened natural systems are frequently driven by (changes in) climate variability that manifests itself in the form of extreme weather events. Combining this with observed and anticipated impacts on unique and threatened human settlements, particularly for the Arctic region of the United States, it follows that this column should, in its progression from yellow to red, at best parallel that for risks of extreme weather events.

The color scheme for aggregate net damages (column 4) is informed by at least two aggregate economic metrics. The first, illustrated by the calibration of the RICE integrated assessment model portrayed by Nordhaus and Boyer (2000), places the net economic cost of climate change for the United States associated with a 4.5°F (2.5°C) warming relative to 1990 levels at 0.45 percent of market gross domestic product (GDP). This estimate includes a willingness to pay of 0.44 percent of market GDP to eliminate a 1.2 percent chance that a permanent loss of 25 percent of global economic income might occur (a reflection of the damage associated with uncertain catastrophic loss) as well as more modest net costs in agriculture and energy. The second metric reflects calculations of annual contributions to the social cost of carbon[1] for the United States alone along alternative baselines with a range of assumed climate sensitivity. Tol and Anthoff (2008) supplied such estimates—derived from the FUND integrated assessment model for the aggregate and for a collection of sectors—to the Environmental Protection Agency. Even for trajectories characterized by high climate sensitivities, none of their (undiscounted) aggregate estimates peak above $3.50 (2000$) per ton of carbon, and most fall short of $1.00 per ton of carbon. Meanwhile, the few sectoral contributions to the aggregate that climb over time do not accelerate until late in the century. Because this time frame would put the increase in global mean temperature above 1990 levels in the 2.7°F–9.9°F (1.5°C–5.5°C) range, the column for aggregate net damages depicted in Figure 2.9 turns from white to light yellow around 3.6°F (2°C). It also reflects both the Nordhaus and Boyer aggregate estimates and the Tol and Anthoff trajectories by not turning orange until global mean temperatures increase by nearly 5.4°F (3°C) and not turning red they pass above 5.4°F (3°C).

Many, if not all, of the risks associated with extreme weather events have asymmetric impacts; this asymmetry is clearly the source of unevenly distributed vulnerabilities that are reflected in column 4, devoted to the distribution of impacts. Even though it will be argued that aggregate economic indicators do not appear to be very sensitive to increases in global mean temperature below 5.4°F (3°C) or so, this diversity of impacts is reflected in the color shading of this column. It is important to recognize

---

[1] The social cost of carbon estimates the discounted economic damages associated with the emission of an extra tonne of carbon at any point in time. It is highly dependent on the future scenario of emissions and development, on a variety of preference parameters like the time preference and aversion to inequality or risk, and a collection of scientific parameters like climate sensitivity. It must be emphasized, as well, that the estimates quoted here are for the United States alone; that is, they do NOT include estimates of economic contributions to the social cost of carbon from damages felt beyond our borders.

that vulnerability to sea level rise is, for example, hardly ever generated by sea level rise itself. Coastal vulnerability is, instead, increased by even modest amounts of sea level rise due to the character of coastal storms whose impacts are very local and distributed differently along the coast, and it is this characteristic that carries an important lesson about the sources of risk. New York City has begun to consider the threat that coastal storms might pose to its vital infrastructure (see Chapter 4). Planners there are beginning to consider investing billions of dollars in projects that are designed primarily to protect vital infrastructure from flooding associated with extreme weather (coastal storms and simply extreme precipitation events). On the opposite coast, planners in California have already engineered a $13 billion protection project for San Francisco Bay (Moser et al., 2009). They have not implemented the requisite investment, but it is estimated that it would involve annual maintenance expenditures in excess of $1 billion per year after construction. If implemented in time, it is likely to save infrastructure worth 10 times the original investment (Moser et al., 2009).

The fundamental point illustrated here is that the distributional impacts of climate change that are buried in the aggregation required to produce national economic estimates are driven to a large degree by the incidence of extreme events and the capacity of specific communities, subcommunities, or systems to respond. Because the column for extreme events misses this adaptation component, the color progression here proceeds less rapidly. It begins at yellow because dramatically asymmetric distributional impacts have already been observed, and it changes quickly to orange because the implicit equity implications of extreme events must also be recognized. As demonstrated by Kates et al. (2006), the poor, the elderly, and perhaps the ethnically disadvantaged are the most vulnerable because of high exposure and high sensitivity. Red shading begins to appear around 3.6°F (2°C). Many of these risks may not appear in the aggregate economic estimates, but they begin to pile up below 3.6°F (2°C). By virtue of their diversity and collective coverage, turning to red then reflects the view that disparate regional indicators of concern should perhaps be used as a national indicator of risk calibrated in noneconomic metrics.

IPCC (2007a) reported on a number of potential futures that would involve large-scale and possibly abrupt climate change. NSTC (2008) briefly discussed ice-sheet contributions to global sea level rise and the chance of significant weakening of the meridional overturning circulation (MOC). Smith et al. (2009a) amplified these assessments by reporting that the risk of additional contributions to sea level rise from both the Greenland and possibly the Antarctic ice sheets may be larger than projected by ice-sheet models and could occur over shorter time scales. This could cause an additional contribution to sea level rise of more than 4 meters, and the climate system could be committed to that future with an increase in global mean temperature of about 3.6°F

(2°C) above 1990 levels. Smith et al. (2009a) also noted increased confidence in projections of carbon cycle feedbacks with potentially far-reaching consequences. In addition, these sources report results from Challenor et al. (2006) that some newer models suggest it is possible that as little as 4.5°F (2.5°C) of additional warming could possibly commit the planet to a significant MOC weakening and/or collapse. Because any of these sources of abrupt change would affect the United States as much as anywhere else, the column for risks of large-scale discontinuities (column 5) depicted in Figure 2.9 duplicates the reasons for concern for the globe depicted by Smith et al. (2009a).

The last column of the figure refers to the National Security Concern for the United States. A report released by the Military Advisory Board (MAB, 2007) summarizes the results of a relatively thorough review of security concerns for the United States that are derived from observed and prospective manifestations of climate change around the world; it is illustrative of the documents from the military and intelligence communities summarized in Chapter 6. Specific impacts extracted largely from IPCC (2007b) for Asia, Africa, South America, Europe, and the Arctic attracted the MAB's attention. The MAB viewed these impacts and vulnerabilities through a risk-management lens that revealed significant risks of social upheaval around the world (e.g., from migration pressures and humanitarian crises in the wake of extreme events like floods, droughts, and severe coastal storms). Because most of the evidence that supported the report's findings was derived from "Risks of Extreme Weather Events" distributed across the globe and well into the long-term future, the global column from Smith et al. (2009a) is replicated in Figure 2.9 as a representation of the sensitivity of national security concerns to changes in global mean temperature.

Acronyms and Initialisms

AAAS	American Association for the Advancement of Science
AAG	Adaptation Assessment Guidebook
AIACC	Assessments of Impacts and Adaptations to Climate Change
ASCE	American Society of Civil Engineers
ASFPM	Association of State Floodplain Managers
BCA	benefit-cost analysis
CCAP	Center for Clean Air Policy
CIG	Climate Impacts Group
CILA	Comision Internaciónal de Límites y Aguas
CNRA	California Natural Resources Agency
CO_2	carbon dioxide
COAG	Council of Australian Governments
COP	Conference of the Parties
CRC	U.S. Climate Resilient Communities
CZMA	Coastal Zone Management Act of 1972
DA	decision analysis
DOT	Department of Transportation
ENSO	El Niño-Southern Oscillation
EPA	Environmental Protection Agency
FEMA	Federal Emergency Management Agency
GAO	Government Accountability Office
GCM	global climate model
GEF	Global Environment Fund
GHG	greenhouse gas
GYE	Greater Yellowstone Ecosystem
HMGP	Hazard Mitigation Grant Program
IAW	Immediate Action Workgroup
IBWC	U.S. International Boundary and Water Commission
ICLEI	ICLEI-Local Governments for Sustainability
IPCC	Intergovernmental Panel on Climate Change
LDCF	Least Developed Countries Fund
LECZ	low elevation coastal zone
MCA	multicriteria analysis
MEA	Millennium Ecosystem Assessment

MME	multimodel ensemble
MPO	metropolitan planning organization
NAPA	National Adaptation Plan for Action
NFIP	National Flood Insurance Program
NGA	National Governors Association
NGO	nongovernmental organization
NOAA	National Oceanic and Atmospheric Administration
NPCC	New York Panel on Climate Change
NRC	National Research Council
NROC	Northeast Regional Ocean Council
NSF	National Science Foundation
OFDA	Office of U.S. Foreign Disaster Assistance
OSHA	Occupational Safety and Health Administration
PDM	Pre-Disaster Mitigation Grant Program
PPE	perturbed physics ensemble
PWWS	Philadelphia Hot Weather–Health Watch/Warning System
RISA	Regional Integrated Sciences and Assessment
SAP	Synthesis and Assessment Product
SARA	Superfund Amendments and Reauthorization Act
SERC	State Emergency Response Commission
UKCIP	United Kingdom Climate Impacts Program
UNDP	United Nations Development Program
UNEP	United Nations Environmental Program
UNFCCC	United Nations Framework Convention on Climate Change
USACE	U.S. Army Corps of Engineers
USAID	U.S. Agency for International Development
USDA	U.S. Department of Agriculture
USGCRP	U.S. Global Change Research Program
USGS	U.S. Geological Survey
WRE	Wigley et al., 1996